AF531504

GREEN EDUCATION

About the Authors

Mahendra K. Satapathy, is presently working as Professor in Science Education and Dean Academic at Regional Institute of Education (NCERT), Bhubaneswar. After passing M.Sc. (Botany) from Utkal University, Bhubaneswar with first rank and university gold medal he completed his Ph.D from ICAR-National Rice Research Institute (NRRI) Cuttack. Having worked as Senior Post-Doctoral Fellow at International Rice Research Institute (IRRI), Manila, Philippines, British Council Fellow at Wye College (University of London) and a Fulbright Fellow in University of Minnesota, USA, he has been working as Professor in National Council for Educational Research and Training (NCERT) New Delhi for more than last ten years. With a Teaching and Research Career of 32 years in Post Graduate Department of Life Science, Dr. Satapathy has headed the Departments of Science and Mathematics and Department of Extension Education, besides being felicitated as the Best teacher of Botany in Odisha. Prof. Satapathy has published a large number of Research Papers and supervised a couple of students for Ph.D in Botany, Life Science and Education. His fields of interest include Science Education, Environmental Education and Sustainability besides Educational Innovations in the country.

Sidhanta Sekhar Bisoi, is presently working as Junior Research Fellow in the Department of Botany, Regional Institute of Education (NCERT), Bhubaneswar, Having completed his M.Sc. and M.Phil. in the field of Biodiversity and Conservation of Natural Resources from Central University of Orissa, Koraput. He has published a series of research papers in International journals of repute and presented his research works in a number of National and International conferences/seminars. His area of interest includes Ethno-botany, Plant diversity studies, and Biodiversity Conservation. Currently, Mr. Bisoi is working in the area of Environmental Sustainability and Livelihood promotion.

GREEN EDUCATION
Plants for Fun and Games

M.K. Satapathy
Sidhanta Sekhar Bisoi
Department of Education in Science & Mathematics
Regional Institute of Education (NCERT)
Bhubaneswar - 751022, Odisha, India

NEW INDIA PUBLISHING AGENCY
New Delhi – 110 034

NEW INDIA PUBLISHING AGENCY
101, Vikas Surya Plaza, CU Block, LSC Market
Pitam Pura, New Delhi 110 034, India
Phone: + 91 (11) 27 34 17 17 Fax: + 91 (11) 27 34 16 16
Email: info@nipabooks.com
Web: www.nipabooks.com

Feedback at feedbacks@nipabooks.com

ISBN: 978-93-87973-08-4

Composed and Designed by NIPA

Dr. Achyta Samanta
Founder, KIIT & KISS
General President, ISCA-2017

Foreword

One of the major concerns today round the world is for conservation of natural resources especially the Biodiversity, the varieties of plants and other living organisms that we see and know. In earlier days children used to play with plants around them and formed an integral part of nature. They design various innovative games and play with stems, fruits, flowers, seeds etc. of the plants and make fun of those.

However, with urbanization and shrinking of wastelands and play grounds, children are putting their leisure time with electronic gadgets such as computers, mobile phones, video games etc. and their traditional/indigenous knowledge associated with nature is getting eroded.

In the present book '**Green Education: Plants for Fun and Games**', the authors have collected this vanishing indigenous knowledge from diverse rural and tribal pockets of Odisha and have described 90 plant species which parts such as stem, fruits, flowers, leaves, seeds etc. have been used for fun and games supplemented with pictures, diagrams and photographs.

I am sure that this book shall be useful to students, teachers, botanists and environmentalists especially the plant lovers as a whole. By the way, I take this opportunity to thank Prof. Mahendra K. Satapathy (Dean Academic) and Mr. Sidhanta S. Bisoi (Research Fellow), Regional Institute of Education (NCERT), Bhubaneswar, for their unique efforts in putting a large amount of information in simple readable style for the benefit of the readers.

Achyuta Samanta

Preface

Currently one of the issues that has attracted the attention of the people all over the world, is the curiosity to know the varieties of plants and animals residing on the surface of the planet earth. This diversity of living creatures including microorganisms has been utilized by each civilization for growth and development as a support system. The rich biodiversity has not only been instrumental in providing humanity with food security, shelter and health care but to a great extent industrial goods leading to high quality of life in the modern world. Further biodiversity is the very stuff that supports the evolution and differentiation of various species over time . It sustains the bodies we live in and affects the lives we lead and the societies we form

Biodiversity also makes irreplaceable contribution to our aesthetics, imagination and creativity. People all over the world visit national parks, sanctuaries and resorts to recreate themselves. It not only helps them to distress but also helps them to feel one with nature.

Teaching is the exploration of creative spirit of Joy. Earlier children considered nature as their teacher. They were using trees, their parts and other natural resources such as soil, sand, pebbles etc as their learning materials and sometimes toys to play. People from their early childhood had constant interaction with nature, thus do make fun of it, respect it, and in turn nature always kept showering its blessings. They use their imagination with plant products such as fruits, flowers. leaves *etc*. to make innovative games to play.

But today, the scenario is different. With urbanization and developments in Science and Technology, present generation children are more curious for electronic gazettes and often engaged in computers, mobile phones, internets, videogames etc rather than gardens and parks as they keep themselves away from nature, they have little or no concern for it. The indigenous knowledge gathered over the years is getting eroded slowly. Interestingly UNO has declared 2011–2020 as the decade on Biodiversity Conservation in order to create awareness about Biodiversity, its importance and conservation.

The present book **"Green Education: Plants for Fun and Games",** based on wide survey in different parts of Odisha, highlights 90 plants species which parts such as fruit, seed, leaf, flower stem etc, are used as play material by children. Besides describing the plant details along with its local name and distribution, the nature of the game or fun has been described with relevant pictures and diagrams. The possible use of the plant part in the teaching- learning process has been highlighted. The authors have also described the economic value of the concerned plant species wherever possible. The book will go a long way to create love for plants, promote aesthetic values and help to conserve this indigenous knowledge.

The book will be useful to plant lovers besides students and teachers of Botany, Life Science and Environmental Studies.

Authors

Contents

Foreword .... *v*

Preface .... *vii*

1. *Abelmoschus esculentus* (L.) Moench 1

English Name: Lady's Finger

Hindi Name: Bhindi

Habit & Distribution 1

General Characters 1

Part Used as Play Material 1

2. *Abrus precatorius* L. 3

English Name: Indian liquorice

Hindi Name: Ratti/Gunchi

Habit & Distribution 3

General Characters 3

Part Used as Play Material 3

3. *Abutilon indicum* (L.) Sweet 5

English Name: Indian mallow

Hindi Name: Kanghi

Habit & Distribution 5

General Characters 5

Part Used as Play Material 5

4. *Acalypha hispida* Burm.f. 7

English Name: Red hot cat's tail

Odia Name: Bilei lanja

Habit & Distribution 7

General Characters 7

Part Used as Play Material 7

5. *Achyranthes aspera* L. 9
English Name: Prickly chaff flower
Hindi Name: Chirchita
Habit & Distribution 9
General Characters 9
Part Used as Play Material 9

6. *Adenanthera pavonina* L. 11
English Name: Red Lucky Seed
Hindi Name: Badi gumchi
Habit & Distribution 11
General Characters 11
Part Used as Play Material 11

7. *Aeschynomene aspera* L. 13
English Name: Pith plant
Hindi Name: Sola
Habit & Distribution 13
General Characters 13
Part Used as Play Material 13

8. *Aegle marmelos* (L.) Correa 15
English Name: Wood apple
Hindi Name: Bel
Habit & Distribution 15
General Characters 15
Part Used as Play Material 15

9. *Albizia saman* (Jacq.) Merr. 17
English Name: Rain Tree
Hindi Name: Gulabi Siris
Habit & Distribution 17
General Characters 17
Part Used as Play Material 17

10. *Allium cepa* L. 19
English Name: Onion
Hindi Name: Pyaz
Habit & Distribution 19

General Characters 19
Part Used as Play Material 19

11. *Areca catechu* L. 21
English Name: Betelnut
Hindi Name: Supari
Habit & Distribution 21
General Character 21
Part Used as Play Material 21

12. *Aristida setacea* Retz. 23
English: Wire grass/spread grass
Odia Name: Suan
Habit & Distribution 23
General Characters 23
Part Used as Play Material 23

13. *Artocarpus heterophyllus* Lam. 25
English: Jack Fruits
Hindi: Kathal
Habit & Distribution 25
General Characters 25
Part Used as Play Material 25

14. *Azadirachta indica* A. Juss: 27
English Name: Neem tree
Hindi Name: Neem
Habit & Distribution 27
General Charcter 27
Part Used as Play Material 27

15. *Dendrocalamus strictus* (Roxb.) Nees 29
English Name: Bamboo
Hindi Name: Bans
Habitat & Distribution 29
General Characters 29
Part Used as Play Material 29

16. *Barleria prionitis* L. 31

English Name: Percupine Flower

Hindi Name: Jhin

Habit & Distribution 31

General Characters 31

Part Used as Play Material 31

17. *Basella Alba* L. 33

English Name: Indian spinach

Hindi Name: Poi

Habit & Distribution 33

General Characters 33

Part Used as Play Material 33

18. *Bauhinia purpurea* L. 35

English Name: Butterfly tree

Hindi Name: Khairwal

Habit & Distribution 35

General Characters 35

Part Used as Play Material 35

19. *Bixa orellana* L. 37

English Name: Annatto

Hindi Name: Latkan/Kumkum

Habit & Distribution 37

General Characters 37

Part Used as Play Material 37

20. *Bombax ceiba* L. 39

English Name: Red silk cotton

Hindi Name: Shalmali

Habit & Distribution 39

General Characters 39

Part Used as Play Material 39

21. *Borassus flabellier* L. 41

English Name: Palm

Hindi Name: Taad

Habit & Distribution 41

General Character 41
Part Used as Play Material 41

22. *Bryophyllum pinnatum* (Lam.) Oken 43
English Name: Miracle leaf
Hindi Name: Zakhm-haiyal
Habit & Distribution 43
General Characters 43
Part Used as Play Material 43

23. *Butea monosperma* (Lam.) Taub. 45
English Name: Bastard teak
Hindi Name: Palash
Habit & Distribution 45
General Characters 45
Part Used as Play Material 45

24. *Caesalpinia bonduc* (L.) Roxb. 47
English Name: Fever nut/Grey Nicker
Hindi Name: Kat-karanj
Habit & Distribution 47
General Characters 47
Part Used as Play Material 47

25. *Caesalpinia pulcherrima* (L.) Sw. 49
English Name: Peacock flower
Hindi Name: Guletura
Habit & Distribution 49
General Characters 49
Part Used as Play Material 49

26. *Calotropis gigantea* R. Br. 51
English Name: Gaint milk weed
Hindi Name: Akoa
Habit & Distribution 51
General Characters 51
Part Used as Play Material 51

27. *Calotropis procera* (Aiton) Dryand 53

English Name: Rubber bush

Hindi Name: Akada

Habit & Distribution 53

General Characters 53

Part Used as Play Material: 53

28 *Carica papaya* L. 55

English Name: Papaya

Hindi Name: Papaya/Papeeta

Habit & Distribution 55

General Characters 55

Part Used as Play Material 55

29. *Cassia fistula* L. 57

English Name: Golden shower

Hindi Name: Sundaraj

Habit & Distribution 57

General Character 57

Part used as play material 57

30. *Citrus aurantium* L. 59

English Name: Orange

Hindi Name: Santra

Habit & Distribution 59

General Characters 59

Part Used as Play Material 59

31. *Clerodendrum infortunatum* L. 61

English Name: Hill glory bower

Hindi Name: Bhant

Habit & Distribution 61

General Characters 61

Part Used as Play Material 61

32. *Cocos nucifera* L. 63

English Name: Coconut

Hindi Name: Nariyal

Habit & Distribution 63

General Characters 63
Part Used as Play Material 63

33. *Codiaeum variegatum* (L.) Rumph. Ex A. juss 65
English Name: Croton
Hindi Name: Croton
Habit & Distribution 65
General Characters 65
Part Used as Play Material 65

34. *Coix lacryma-jobi* L. 67
English Name: Adlay/Job's tear
Hindi Name: Gurlu
Habit & Distribution 67
General Characters 67
Part Used as Play Material 67

35. *Colocasia esculenta* (L.) Schott 69
English Name: Elephant ear taro
Hindi Name: Kachu/Kachlu
Habit & Distribution 69
General Characters 69
Part Used as Play Material 69

36. *Corchorus capsularis* L. 71
English Name: White jute
Hindi Name: Titapat
Habit & Distribution 71
General Charcaters 71
Part Used as Play Material 71

37. *Cordia dichotoma* G.Forst. 73
English Name: Indian cherry
Hindi Name: Lasora
Habit & Distribution 73
General Characters 73
Part Used as Play Material 73

38. ***Crotalaria pallida*** **Aiton. 75**

English Name: Rattle pods

Hindi Name: San

Habit & Distribution 75

General Characters 75

Part Used as Play Material 75

39. ***Cucumis sativus*** **L. var.** ***hardwickii*** **(Royle) Gabaev 77**

English Name: Wild cucumber

Hindi Name: Khira

Habit & Distribution 77

General Characters 77

Part Used as Play Material 77

40. ***Cuscuta reflexa*** **Roxb. 79**

English Name: Giant dodder

Hindi Name: Amar bel

Habit & Distribution 79

General Characters 79

Part Used as Play Material: Stem (vines) 79

41. ***Cynodon dactylon*** **(L.) Pers. 81**

English Name: Bermuda grass

Hindi Name: Dub

Habit & Distribution 81

General Characters 81

Part Used as Play Material 81

42. ***Datura metel*** **L. 83**

English Name: Thron apple

Hindi Name: Dhatura

Habit & Distribution 83

General Characters 83

Part Used as Play Material 83

43. ***Dalbergia sissoo*** **DC. 85**

English Name: Rose wood

Hindi Name: Shisham

Habit & Distribution 85

General Characters 85
Part Used as Play Material 85

44. *Delonix regia* (Hook.) Raf. 87
English Name: Gold mohar
Hindi Name: Gul mohar
Habit & Distribution 87
General Characters 87
Part Used as Play Material 87

45. *Dillenia indica* L. 89
English Name: Elephant apple
Hindi Name: Chalta
Habit & Distribution 89
General Characters 89
Part Used as Play Material 89

46. *Drypetes roxburghii* (Wall.) Hurus. 91
English Name: Lucky bean tree
Hindi Name: Putija/Putranjiva
Habit & Distribution 91
General Characters 91
Part Used as Play Material 91

47. *Erythrina variegata* L. 93
English Name: Indian coral tree
Hindi Name: Pangara
Habit & Distribution 93
General Characters 93
Part Used as Play Material 93

48. *Eucalyptus globulus* Labill. 95
English Name: Blue gum
Hindi Name: Nilgiri
Habit & Distribution 95
General Characters 95
Part Used as Play Material 95

49. *Ficus benghalensis* L. 97

English Name: Banyan tree

Hindi Name: Barh

Habit & Distribution 97

General Characters 97

Part Used as Play Material 97

50 *Ficus racemosa* L. 99

English Name: Cluster fig tree

Hindi Name: Goolar

Habit & Distribution 99

General Characters 99

Part Used as Play Material 99

51. *Ficus religiosa* L. 101

English Name: Pipal tree

Hindi Name: Peepal

Habit & Distribution 101

General Characters 101

Part Used as Play Material 101

52. *Impatiens balsamina* L. 103

English Name: Garden Balsam

Hindi Name: Gul-mendi

Habit & Distribution 103

General Characters 103

Part Used as Play Material 103

53. *Ipomoea hederifolia* L. 105

English Name: Scarlet morning glory

Hindi Name: Lalpungli

Habit & Distribution 105

General Characters 105

Part Used as Play Material 105

54. *Ipomoea parasitica* (Kunth) G. Don 107

English Name: Yellow-Throated Morning Glory

Hindi Name: Neelkalme

Habit & Distribution 107

General Characters 107
Part Used as Play Material 107

55. *Jasminum arborescens* Roxb. 109
English Name: Royal jasmin
Hindi Name: Chameli
Habit & Distribution 109
General Characters 109
Part Used as Play Material 109

56. *Jatropha gossypiifolia* L. 111
English Name: Bellyache bush
Hindi Name: Ratanjoti
Habit & Distribution 111
General Characters 111

57. *Jatropha integerrima* Jacq. 113
English Name: Peregrina/Spicy jatropha
Habit & Distribution 113
General Characters 113
Part Used as Play Material 113

58. *Lagenaria siceraria* (Molina) Standl. 115
English Name: Calbash
Hindi Name: Lauki
Habit & Distribution 115
General Characters 115
Part Used as Play Material 115

59. *Lantana camara* L. 117
English Name: Wild sage
Hindi Name: Raimuniya
Habit & Distribution 117
General Characters 117
Part Used as Play Material 117

60. *Luffa acutangula* (L.) Roxb. 119
English Name: Angled luffa
Hindi Name: Karvito
Habit & Distribution 119

General Characters 119
Part Used as Play Material 119

61. *Mangifera indica* L. 121

English Name: Mango
Hindi Name: Aam
Habit & Distribution 121
General Characters 121
Plant Part Used as Play Material 121

62. *Martynia annua* L. 123

English Name: Tiger's claw/Devil's claw
Hindi Name: Ulat kanta
Habit & Distribution 123
General Characters 123
Part used as Play Material 123

63. *Mimosa pudica* L. 125

English Name: Touch me not/sensitive plant
Hindi Name: Lajwanti
Habit & Distribution 125
General Characters 125
Part Used as Play Material 125

64. *Mimusops elengi* L. 127

English Name: Spanish cherry
Hindi Name: Maulsari
Habit & Distribution 127
General Characters 127
Part Used as Play Material 127

65. *Musa paradisiaca* L. 129

English Name: Banana
Hindi Name: Kela
Habit & Distribution 129
General Characters 129
Part Used as Play Material 129

66. *Neolamarckia cadamba* (Roxb.) Bosser. 131

English Name: Bur flower

Hindi Name: Kadamb

Habit & Distribution 131

General Characters 131

Part Used as Play Material 131

67. *Nicandra physalodes* (L.) Gaertner 133

English Name: Shoo-fly plant

Hindi Name: Popti

Habit & Distribution 133

General Characters 133

Part used as Play Material 133

68. *Nymphaea pubescens* Willd. 135

English Name: Water lily

Hindi Name: Kanval

Habit & Distribution 135

General Characters 135

Part Used as Play Material 135

69. *Alangium salviifolium* (L.f.) Wangerin 137

English Name: Sage Leaved Alangium

Hindi Name: Ankol

Habit & Distribution 137

General Characters 137

Part Used as Play Material 137

70. *Oryza sativa* L. 139

English Name: Rice

Hindi Name: Chawal

Habit & Distribution 139

General Characters 139

Part Uses as Play Material 139

71. *Pandanus odorifer* (Forssk.) Kuntze 141

English Name: Kewda/Screw pine

Hindi Name: Gagan dhul

Habit & Distribution 141

General Characters 141
Part Used as Play Material 141

72. *Peltophorum pterocarpum* (DC.) K. Heyne 143

English: Yellow flamebouyant
Hindi: Peela gulmohar
Habit & Distribution 143
General Characters 143
Part Used as Play Material 143

73. *Phoenix sylvestris* (L.) Roxb. 145

English Name: Date palm
Hindi Name: Khajur
Habit & Distribution 145
General Characters 145
Part Used as Play Material 145

74. *Plumbago zeylanica* L. 147

English: Wild leadwort
Hindi: Chitrak
Habit & Distribution 147
Gerneral Characters 147
Part Used as Play Material 147

75. *Polyalthia longifolia* (Sonn). Thwaites 149

English Name: False Asoka
Hindi Name: Ashok
Habit & Distribution 149
General Characters 149
Part Used as Play Material 149

76. *Pseudobombax ellipticum* (Kunth) Dugand 151

English Name: Shaving brush tree
Habit & Distribution 151
General characters 151
Part Used as Play Material 151

77. *Ricinus communis* L. **153**
English Name: Castor bean
Hindi Name: Arandi
Habit & Distribution 153
General Characters 153
Part Used as Play Material 153

78. *Rosa indica* L **155**
English Name: Rose
Hindi Name: Gulab
Habit & Distribution 155
General Characters 155
Part Used as Play Material 155

79. *Ruellia tuberosa* L. **157**
English Name: Fever root
Hindi Name: Ruwel
Habit & Distribution 157
General Characters 157
Part Used as Play Material 157

80. *Saccharum officinarum* L. **159**
English Name: Sugarcane
Hindi Name: Ganna
Habit & Distribution 159
General Characters 159
Part Used as Play Material 159

81. *Solanum viarum* Dunal **161**
English Name: Tropical soda apple
Odia Name: Denga bheji/Bheji baigana
Habit & Distribution 161
General Characters 161
Part Used as Play Material 161

82. *Tagetes erecta* L. **163**
English Name: Marigold
Hindi Name: Genda
Habit & Distribution 163

General Characters 163
Part Used as Play Material 163

83. *Tamarindus indica* L. 165

English Name: Tamarind
Hindi Name: Imli
Habit & Distribution 165
General Characters 165
Part Used as Play Material 165

84. *Tectona grandis* L.f. 167

English Name: Teak
Hindi Name: Sagun
Habit & Distribution 167
General Characters 167
Part Used as Play Material 167

85. *Tinospora cordifolia* (Willd.) Miers. 169

English Name: Indian Tinospora
Hindi Name: Gulbel
Habit & Distribution 169
General Characters 169
Part Used as Play Material 169

86. *Trewia nudiflora* L. 171

English Name: False white teak
Hindi Name: Pindar
Habit & Distribution 171
General Characters 171
Part Used as Play Material 171

87. *Tridax procumbens* (L.) 173

English Name: Coat buttons
Hindi Name: Khal-Muriya
Habit & Distribution 173
General Characters 173
Part Used as Play Material 173

88. *Triumfetta rhomboidea* Jacq. **175**

English Name: Burr Bush

Hindi Name: Chikti

Habit & Distribution 175

General Characters 175

Part Used as Play Material 175

89. *Combretum indicum* (L.) Defilipps **177**

English Name: Rangoon creeper

Hindi Name: Madhumalti

Habit & Distribution 177

General Characters 177

Part Used as Play Material 177

90. *Xanthium strumarium* L. **179**

English Name: Common Cocklebur

Hindi Name: Chhota dhatura/Ghagra

Habit & Distribution 179

General Characters 179

Part Used as Play Material 179

List of Plants and their Parts Used for Fun and Games **181**

1

Abelmoschus esculentus (L.) Moench

Common name: English – Lady's finger

Hindi – Bhindi

Odia- Bhendi

Family: Malvaceae

Habit & Distribution

The plant is a herb (Fig. 1 a), widely cultivated as vegetable in the tropical countries including India.

General Characters

It is an annual plant, grows about 0.5-2 m in height. Leaves are orbicular, 5-25 cm in diameter, 5-7 lobed or variously dissected and cordate. The flowers are about 5 cm in diameter, pedicel 5-15 cm with scattered simple hairs. The corolla of the flower is yellow or cream in colour with dark purple at the centre. Fruit is an elongated capsule with a pointed end. The plant bears flowers and fruits throughout the year.

Part Used as Play Material

The fruit is usually used as vegetable for cooking purpose. Before cooking the stalk of the fruit is removed. These left out stalks are often used by children as horns putting those on their foreheads (Fig. 1 b). The basal rounded part carrying mucilage gets attached to the forehead and children enjoy it.

Fig. 1: (a) *Abelmoschus esculentus* (L.) Moench

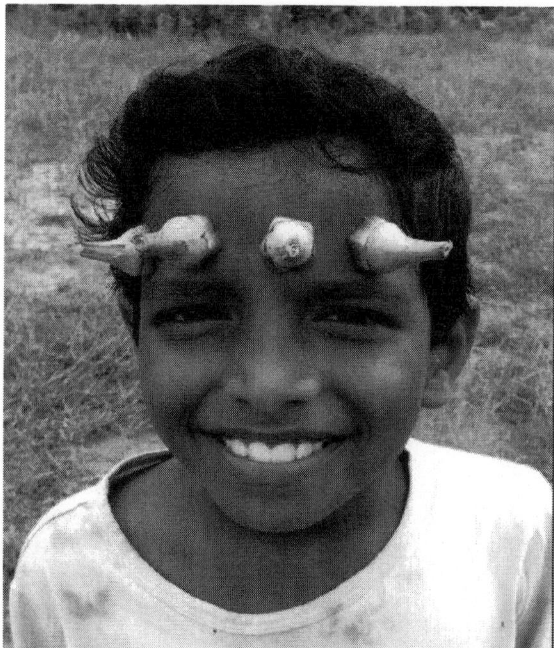

Fig. 1: (b) Children playing with fruit stalks as Horns

2

Abrus precatorius L.

Common name: English- Indian liquorice

Hindi- Ratti/Gunchi

Odia- Kaincha

Family : Fabaceae

Habit & Distribution

It is a common weed (Fig. 2 a) found in forest and it is also seen growing on road sides, gardens, waste places and sometimes planted in gardens.

General Characters

The plant is a slender, perennial climber that twins around trees. It has no special organ of attachment. It has slender branches and a cylindrical wrinkled stem. Leaves are glabrous with long internodes, alternate, compound parpinnate with stipules. Each leaf is about 50-100 mm long, having 20-40 leaflets. Flowers are small, pale violate with short stalk, arranged in clusters. The fruit is a pod, which is flat, oblong, and truncate shaped with silky texture. Pod contains 3-5 oval shaped seeds. They are usually bright scarlet with smooth, glossy texture and a black patch on top.

Part Used as Play Material

Girls use seeds for making necklace because of its attractive color. The seed of the plant is deep red in color having a black spot on its tip. So, it looks very attractive and gorgeous, for which girls use it for making necklace by joining the seeds in series by thread. Children are also used it to make bracelets and other models for their fun sake (Fig.2 b).

Fig. 2 (a) *Abrus precatorius* L.

Fig. 2 (b) Girls use seeds for making necklace

3

Abutilon indicum (L.) Sweet

Common name : English- Indian mallow

Hindi- Kanghi

Odia- Pedipedica

Family : Malvaceae

Habit & Distribution

It is a shrub (Fig.3 a) commonly found in waste places, along road sides, as weeds in garden and in forests. It is distributed throughout India.

General Characters

The plant is an erect velvety-pubescent shrub with circular or heart shaped leaves and grows up to 1-2 m height. The leaves are alternately arranged and have long stalk with velvety, soft and pale hairs. Flowers are orange-yellow in colour, 2-3 cm across and occur solitarily in axils. Fruits are circular in shape, consisting of 11-20 radiating hairy carpels, looks brown when dry. Seeds are kidney shaped. The plant flowers and bears fruits during July-April.

Part Used as Play Material

The fruits are circular in shape consisting of radiating hairy carpels on it. It looks rounded and star shaped. Children use these fruits as their play material by joining a number of fruits in a series by a thread (Fig.3b) that looks like a tiffin carrier.

Fig. 3(a): *Abutilon indicum* (L) Sweet

Fig. 3(b): Fruit are used in funmaking by joining in series

4

Acalypha hispida Burm.f.

Common name : English- Red hot cat's tail

Odia- Bilei lanja

Family : Euphorbiaceae

Habit & Distribution

The plant usually seen along the road sides and often grown in gardens (fig 4 a).

General Characters

It is a bushy shrub grows up to 1.8 to 3.7 meters in height. The leaves are ever green, oval, 10-23 cm long and 7.5-10 cm in width with pointed ends. Flowers are borne in attractive shades, ranging in colour from deep purple to bright red and clustered in velvety catkins, which are about 30-50 cm long. The plant flowers throughout the year.

Part Used for Play Material

The inflorescences is long and deep red in colour that looks attractive and appears like a cat's tail. Children play with the flowers by putting it as a tail in the back of their pants (Fig.4 b). Also girls used it to decorate their hands as ornaments.

Fig. 4 (a) ***Acalypha hispida*** **Burm.f.**

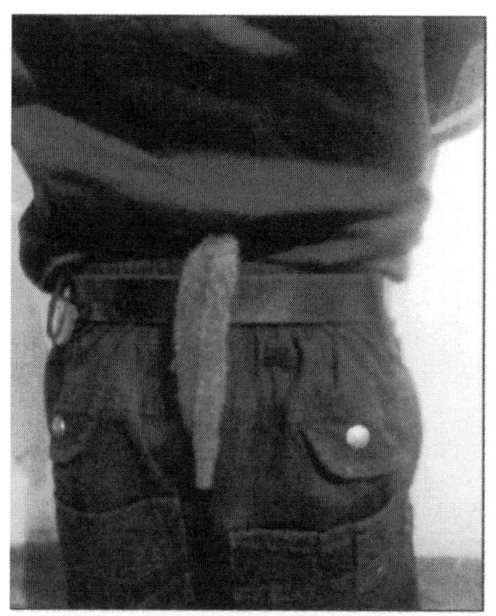

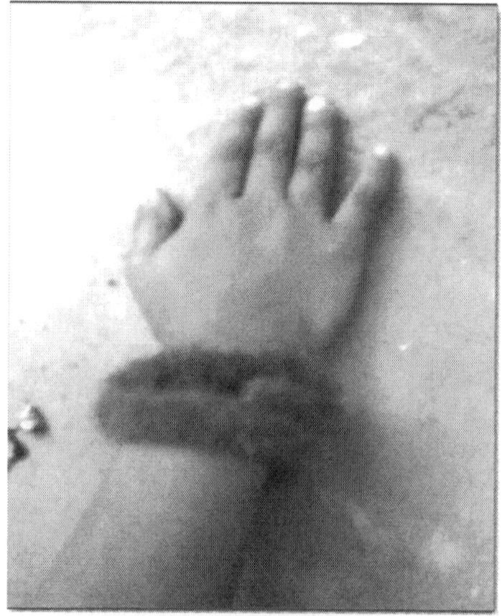

Fig. 4 (b) Children use the flower as tail and also as ornaments

5

Achyranthes aspera L.

Common name : English- Prickly chaff flower

Hindi- Chirchita

Odia- Apamaranga

Family : Amaranthaceae

Habit & Distribution

The plant is a herb (Fig 5 a), fairly common weed that grows on waste lands and distributed throughout India.

General Characters

The plant is an erect, straggling or subscandant herb. The leaves are ovate, elliptic, obtuse, acute and finely attached to the stem on both sides. The flowers are greenish white in colour, bracteolate having membranous blade. Seeds are reddish brown in colour, subcylindric at apex and rounded at the base. The flowering and fruiting of the plant occurs during the month October-February.

Part Used as Play Material

The leaves of the plant are very thin. Children use the leaves as cracker by folding the leaves inside palm (Fig.5 b).

The plant have also medicinal value and used against vomiting and dysentery.

Fig. 5 (a) *Achyranthes aspera* L.

Fig. 5 (b) Children press the leaves as cracker

6

Adenanthera pavonina L.

Common name : English- Red Lucky Seed

Hindi- Badi gumchi

Odia- Manda kaincha

Family : Mimosaceae

Habit & Distribution

The plant is a tree (Fig. 6a) often found in gardens and village sides and distributed in different states of India.

General Characters:

It is a large tree with grey bark. The leaves are compound bipinnate, green when young, turning yellow when old and having 5-6 pairs of leaflets. The flowers are yellowish, small and grow in dense drooping and about 5-15 cm long. Fruits are curved, hanging, green pods that turn brown, coil up and split open as they ripen to reveal small bright red seeds. The plant flowers during March-May.

Part Used as Play Material

The seeds are rounded and deep red in colour that looks very attractive. Children love these red seeds and usually collect the seeds littered under the tree. They use the seeds to make different play materials like necklace and bracelets for playing games and making fun (Fig.6 b).

Fig. 6 (a) *Adenanthera pavonina* L.

Fig. 6 (b) Necklace is prepared from the seeds as play material

7

Aeschynomene aspera L.

Common name : English- Pith plant

Hindi- Sola

Odia- Sola

Family : Fabaceae

Habit & Distribution

It is a herb (Fig.7a), commonly found in ponds, swampy areas, lakes and other water bodies. It is distributed throughout India.

General Characters

The plant is an erect, perennial herb. Its stem is about 2 cm in diameter. The leaves are sessile reaching up to 15 cm long, leaflets 25-50 pairs, alternate or opposite. The flowers are yellow in colour and about 8 cm long, broadly ovate and acute. The plant flowers between August and October.

Part Used as Play Material

The plant stem has the lightest wood of all. Children use to make boats by arranging the stems in series. The stem is also used to make different crafts and toys as play material, besides being used for decoration purposes and making caps (Fig.7 b).

Fig. 7 (a) *Aeschynomene aspera* L.

Fig. 7 (b) Stem is used to make different crafts and toys as play material

8

Aegle marmelos (L.) Correa

Common name : English- Wood apple

Hindi- Bel

Odia- Bela

Family : Rutaceae

Habit & Distribution

It is a tree, (Fig 8 a) common in scrub and open forests, also planted near temples and villages. The plant is usually distributed throughout the greater part of India.

General Characters

The plant is a medium sized tree with spines of 1-2 cm long. The leaf is composed of three leaflets. The leaflets are oval, elliptic with 2.5-6.5 cm petiole, terminal one is larger. Flowers of the tree are white or greenish-white. The fruits are ovoid, hard and about 5-7.5 cm in diameter. The flowers appear during March-April and the tree bears fruits towards October, ripening in April.

Part Used as Play Material

The fruit of the plant is very hard and woody from outside. Children use it as shortfoot for throwing it from a point (Fig.8 b). Also the unripe fruits are used as ball for catch and throw games as it appears like a playing ball. The fruit carries lots of nutritive value.

The leaves have medicinal and religious importance. It is used for treatment of stomach disorders and for promoting digestion. It also used as holy leaf for different festivals and pujas, especially to worship Lord Shiva, the king of gods.

Fig. 8 (a) *Aegle marmelos* L. Correa

Fig. 8 (b) Children throw fruits as short foot for playing

9

Albizia saman (Jacq.) Merr.

Common name : English- Rain Tree

Hindi- Gulabi Siris

Odia- Chakunda

Family : Mimosaceae

Habit & Distribution

The plant is a tree common in dry, lowlands ,grasslands, coastal bush lands, forests and also planted in avenues. It is distributed throughout the tropical countries (Fig 9 a).

General Characters

The tree is easily recognized by its umbrella like canopy of evergreen, feathery foliage and puffs of pink flowers. It is a large, fast-growing, deciduous tree reaching up to 30 m height. The leaves are alternate, bipinnately-compound, with small asymmetrical leaflets that are more curved on outer margin, and finely velvety on underside. Leaflets fold up in early evening and during overcast days. The Flowers are produced as dense cluster with numerous filamentous stamens, pink above and white below, slightly fragrant. The fruit is a fleshy pod, sweet in taste and much relished by squirrels, horses and cattle. The plant flowers during September-December.

Part Used as Play Material

The leaves of the plant are rounded and equal in size. Children use the leaves as coins while playing different games like purchase and sale. The leaves being equal sized and a bit hard, these are easily put in pant and shirt pockets by the children (Fig 9 b)

Fig. 9 (a) *Albizia saman* (Jacq.) Merr.

Fig. 9 (b) Children use leaves as coins in different games

10

Allium cepa L.

Common name : English- Onion

Hindi- Pyaz

Odia- Piaja

Family : Amaryllidaceae

Habit & Distribution

It is a herb (Fig. 10 a). The plant is commonly cultivated along the river banks during winter season.

General Characters

The leaves of the plants are cylindric, fistular and look like tubes. Flower of the plant is white or greenish white with a short pedicel. The plant flowers during January- February.

Part Used as Play Material

The leaves are long, cylindrical and hollow. The shape is tubular just like a straw or pipe. Children often use the leaves to suck juice/water by putting one end into their mouth and the other in the juice or water bottle (Fig.10 b).

The plant has got food value, as onion is used regularly as a vegetable or as a spice.

Fig. 10 (a) *Allium cepa* L.

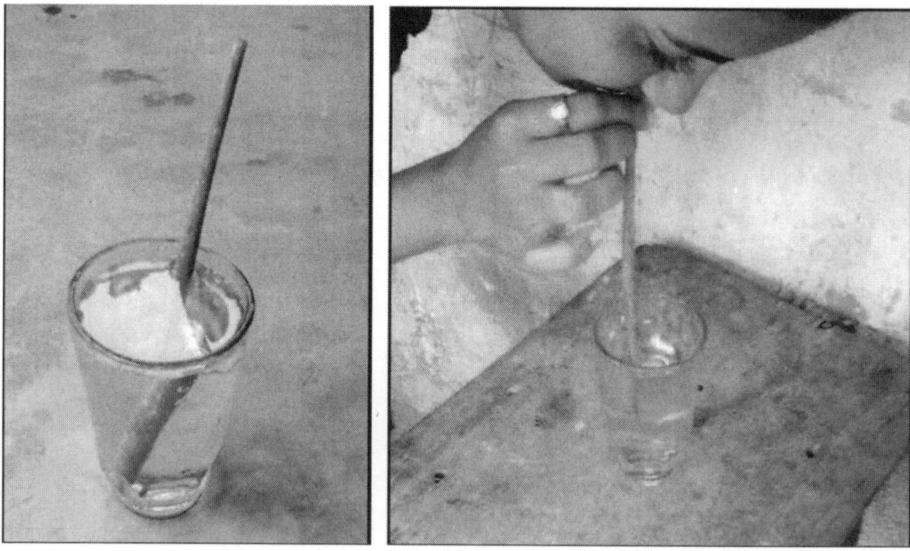

Fig. 10 (b) Children use the leaves to suck juice/water

11

Areca catechu L.

Common name : English- Betelnut

Hindi- Supari

Odia- Gua

Family : Arecaceae

Habit & Distribution

The plant is a tree, (Fig. 11 a) naturalized and commonly cultivated throughout India.

General Character

The plant is perennial, grows up to 20 meters in height and 8-12 inches in diameter. The stem is solitary, slender and erect. The leaf of the plant is 1.5-2 meter long, pinnate with numerous crowded leaflets. Fruit is a coriaceous drupe, ovoid, 3.7-5 cm, smooth orange, with a single seed with truncate base. The plant flowers and bears fruits throughout the year.

Part Used as Play Material

The fruit is oval in shape, and orange in colour and 3.7-5 cm in size, that looks like a small orange ball. Children use the fruits as playing ball to play different games such as guli. Children also use the leaf branch as lorry carrier (Fig.11 b).

The seed has economic value, chewed along with betel leaves mostly in east and north-eastern states.

Fig. 11 (a) *Areca catechu* L.

Fig. 11 (b) Fruit is used as ball and leaf branch as lorry carrier

12

Aristida setacea Retz.

Common name : English - Wire grass/spread grass

Odia- Suan

Family : Poaceae

Habit & Distribution

It is a herb (Fig. 12 a) widely found in agricultural lands, hill sides, road sides, river bank etc. It is distributed throughout India.

General Characters

The plant ascending to erect, grows up to 3-4 feet in height. The leaves may be flat or rolled. The inflorescence either panicle like or raceme with spiky branches. Flowers are tiny and hairs like awns seen protruding from each flower. The plant flowers during September-November.

Part Used as Play Material

The flower is easily removed from the plant and is used to make toy snake like structure from the flower stalk, as play material by the children (Fig.12 b).

Fig 12 (a) *Aristida setacea* Retz.

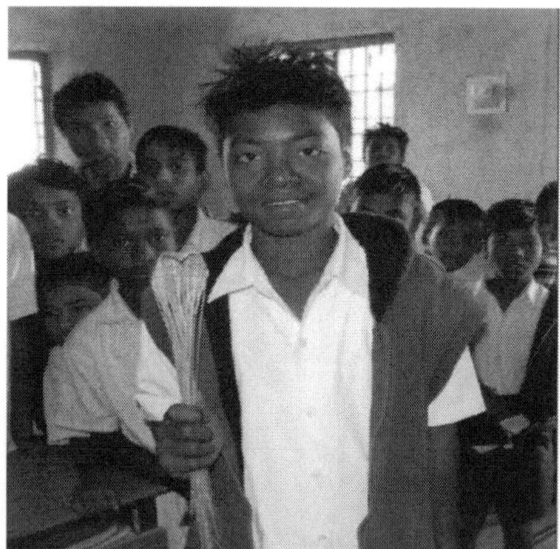

Fig 12 (b) Flowers stalk is used for making snake like toys

13

Artocarpus heterophyllus Lam.

Common name : English- Jack Fruits

Hindi- Kathal

Odia- Panasa

Family : Moraceae

Habit & Distribution

It is a tree (Fig. 13 a) found in wild as well as cultivated in farms. It was introduced and has become naturalized in many parts of India.

General Characters

The plant is an evergreen tree that grows up to 20-30 meter tall and 80-200 cm in diameter with evergreen, alternate, glossy, leathery leaves reaching up to 9 inch long. Leaves are oval in mature trees. All parts of the plant contain sticky, white latex. Inflorescence is solitary, borne axillary on special lateral short leafy shoots arising from older branches. Fruits are large arising from the tree trunk and weighting about 10-50 kg. the exterior of the fruit is green or yellow when ripe and composed of numerous hard, cone like points attached to a thick and rubbery, pale yellow or whitish wall. The plant flowers during December-February and bears fruits during June-July.

Part Used as Play Material

The leaves of the plant are oval, waxy and leathery. Children use these leaves to make utensils like plates, spoons, bowls etc. as their play material for kitchen set game. They also use the leaves for making whistle by rolling it and chakries by cutting it in to different shapes.. Different models of Tiara caps are also prepared by children after joining a number of leaves through coconut stick.

Early fallen fruits are also used by the children to make toys of horse/cow as their play material (Fig. 13 b). The fruits have a lot of nutritive value. It is consumed as raw vegetable and fruit after ripening.

Fig. 13 (a) *Artocarpus heterophyllus* Lam.

Fig. 13 (b) Children use Leaves for making utencils, whistles, tiaras and fruits are used for making toys like cow, horse etc.

14

Azadirachta indica A. Juss

Common name : English- Neem tree

Hindi- Neem

Odia- Nimba

Family : Meliaceae

Habit & Distribution

The plant is a tree (Fig. 14 a), very commonly planted and often self shown. Also seen in forests as well as in wild. It is cultivated and naturalized throughout India.

General Charcter

The leaf of the plant is about 20-38 cm in length, leaflets 5-9 pairs and obliguely lanceolate. Flower is white in clolour and fragrant. The plant bark, leaves and flowers are bitter in taste. The plant bears flowers during February to May. The plant fruiting occurs during June to July.

Part Used as Play Material

Fruit: the fruits, when ripe is 1.4-2.8 cms by 1.0-1.5 cms in size. The fruit skin (exocarp) is thin and the bitter-sweet pulp (mesocarp) is about 0.3-0.5 cm thick. The hard brown colour seeds are imitated as marble pebbles to mark and play by the children.

Stem

The stem is woody and strong. Children use to make bows and arrows by bending these wood pieces from both the ends and binding the ends by a thread, use as play material (Fig.14 b).

The plant has strong medicinal/pesticidal value.

Fig. 14 (a) *Azadirachta indica* A. Juss.

Fig. 14 (b) Children use stem to make bows and arrows

15

Dendrocalamus strictus (Roxb.) Nees

Common name : English- Bamboo

Hindi- Bans

Odia- Baunsa

Family : Gramineae (Poaceae)

Habit & Distribution

It is a perennial tree (Fig. 15a), family common in mixed forest or dry hills, and often found in shady slopes and allovial plains.

General Characters

The plant is a perennial tree grows up to 20-53 m in height and 8-18 cm in diameter. There are 1-3 spines at each branch node. Leaves are thin, linear, up to 20 cm long, 1-1.8 cm wide with a long pointed tip. The branches of the plant spread out from the base and areal roots reaching up to nodes. The length of the internodes is about 15-46 cm and is hollow.

Part Used as Play Material

The stem of the plant is strong, cylindrical and hollow. So, children use the cut pieces of the clums to make glasses and pen stands. Children also use it to make gun, flute, whistle and carts. Often children also use the stem as hockey stick.

The clum sheaths are elongated, plain and have large surface area with light brown colour. Children use it as drum by striking it by sticks to make fun of it (Fig.15 b).

Dendrocalamus strictus (Roxb.) Nees

Fig. 15 (b) Stem is used for making glass, pen stand, gun, flute, whistles etc.

16

Barleria prionitis L.

Common name : English- Percupine Flower

Hindi- Jhinti

Odia- Daskaranta

Family : Acanthaceae

Habit & Distribution

The plant is a shrub (Fig. 16 a), common in waste places, roadsides and dry places. It is distributed throughout India.

General Characters

It is an erect shrub, growing to about 1.5 m tall. The erect leaves are up to 100 mm long and 40 mm wide and oval shaped though narrow at both ends. The flowers are yellow-orange, tubular and found bunched tightly together at the top. The seed capsule is oval-shaped and 13-20 mm long, with a sharp pointed beak. The plant flowers and bears fruits during September-February.

Part Used as Play Material

The fruit of the plant is a pod, that bursts open with a cracking sound when comes in contact with water (Fig.16 b). For making fun children treat the pods as crackers by putting it into water.

Fig. 16 (a) *Barleria prionitis* L.

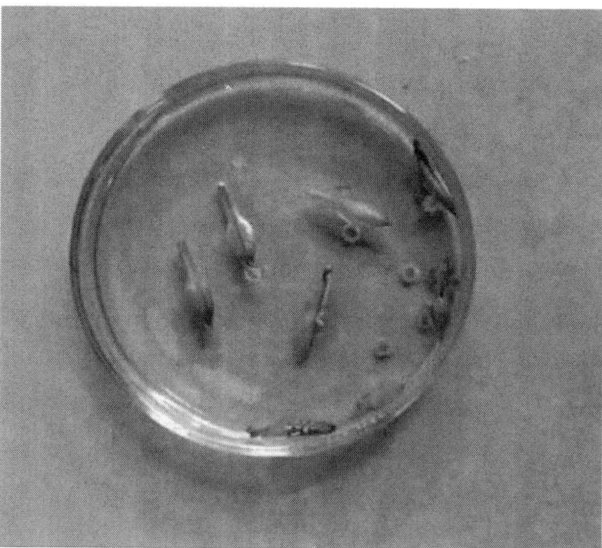

Fig. 16 (b) Children use fruits as crackers by putting it into water

17

Basella Alba L.

Common name : English- Indian spinach

Hindi- Poi

Odia- Poi

Family : Basellaceae

Habit & Distribution

The plant is a twining herb (Fig. 17 a), commonly cultivated and often found in wild state on rocky grassy hill sides. Also widely cultivated throughout India.

General Characters

It is a twining succulent herb with green stem and petioles. Leaves of the plant are broadly ovate and green to purplish in colour. Flowers are pinkish or purple in colour. Fruits are berries, ovoid or depressed-globose, turns black after ripening with deep purple juice.

Part Used as Play Material

Fruits- the fruits are very soft and purple to deep black in colour. There is presence of deep purple juice inside the fruit. For making fun the juice of the fruit is used by children to colour their palms and fingers (Fig.17 b).

The plant has also food value as it is consumed as a vegetable.

Fig. 17 (a) *Basella alba* L.

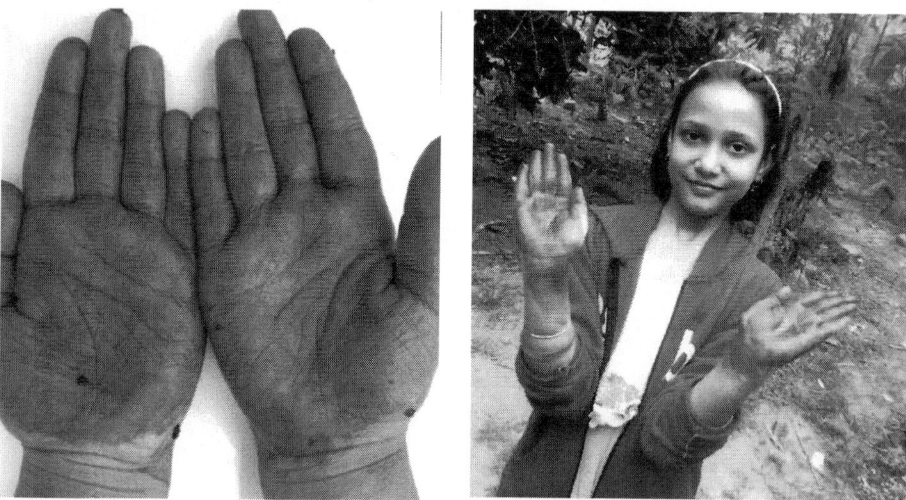

Fig. 17 (b) Fruits are used for coloration of hand

18

Bauhinia purpurea L.

Common name : English- Butterfly tree

Hindi- Khairwal

Odia- Kuilari

Family : Caesalpiniaceae

Habit & Distribution

It is a moderate sized tree (Fig. 18a), seen frequently in forests and often planted. It is distributed in foothill areas also and cultivated throughout India.

General Characters

The plant is a moderate sized tree with ashy or dark brown bark color. Leaves are often oblong, about 5-18, deeply lobed from one third way down to nearly the base with scarcely overlapping. Flowers are purple, large on terminal panicled racemes. Pods are 15-30×18-25 cm long, narrow below, somewhat broadening upwards. The plant flowers during September-February and bears fruits during February-March.

Part Used as Play Material

The leaves of the plant are broad and oblong. Children often use the leaves in play to make containers of different shapes by folding it.

Fig. 18 (a) *Bauhinia purpurea* L.

Fig. 18 (b) Children use Leaves to make Container in diffent games

19

Bixa orellana L.

Common name : English- Annatto

Hindi- Latkan/Kumkum

Odia- Kumkuma

Family : Bixaceae

Habit & Distribution

The plant is a shrub (Fig. 19 a), cultivated and naturalized, also planted in gardens. It is frequently cultivated throughout India.

General Characters

The plant grows up to 6-20 feet tall and survives for about 50 years. Leaves are ovate, large, 5-15 cm long and 4-11 cm wide with pointed ends. Flowers occur in vertical upright clusters. The flowers are showy, white or pink, nearly 5 cm broad. The fruit is 2 valved, ovate with a scarlet covering. The plant flowers during July-October and bears fruits during October-May.

Part Used as Play Material

The fruit of the plant become red when ripe, the seeds present inside the fruit also become deep red in colour. Children use these seeds for colouration of palms and feet by breaking the fruits. It gives a very attractive deep red colour. (Fig. 19 b)

Fig. 19 (a) *Bixa orellana* L.

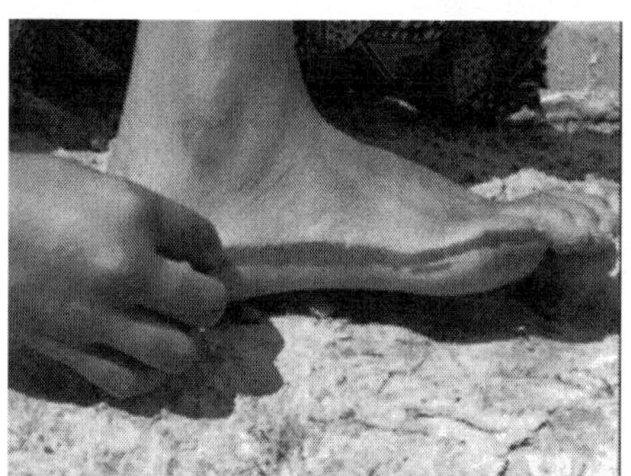

Fig. 19 (b) Seeds are used for colouration of palms and feet

20

Bombax ceiba L.

Common name : English- Red silk cotton

Hindi- Shalmali

Odia- Simili

Family : Bombacaceae

Habit & Distribution:

It is a tree (Fig. 20 a), fairly common throughout India. It is found in hot, dry river valleys, and the deciduous belt of the hills.

General Characters

The plant is a tree with prickly trunk and branches when young and grows up to a height of 25 meters with a spreading crown. Leaves are compound, stipulate, and clustered at twig ends with 12-25 cm long. Flowers are bright red in colour, 10-12 cm across, solitary or in fascicles and usually on leafless branches. Fruits are woody capsule, 5-valved with many seeds. The plant flowers during January-March and bears fruits during March to May.

Part Used as Play Material

The stem is woody and durable. Children use the stem to make different toys like cars, wheels, bullock carts etc. for play purpose. (Fig. 20 b)

Fig. 20 (a) *Bombax ceiba* L.

Fig. 20 (b) Used for making wheels and toys

21

Borassus flabellier L.

Common name : English- Palm

Hindi- Taad

Odia- Tala

Family : Arecaceae

Habit & Distribution

It is a large tree and widely planted crop (Fig. 21a). The plant is native to tropical Africa but cultivated and naturalized throughout india.

General Character

The plant is a large tree and grows upto 30-40 m. in height. The trunk may have a circumference of 1.7 m at the base. The leaves are palmate, leathery, gray green and 1-3 m in wide with folded along at the midrib. Inflorescences are 90-150 cm long and branched. Fruits are ovoid, 15-20 cm in diameter with fibrous and fleshy mesocarp. The plant flowers during February to March and bears fruits during April to August.

Part Used as Play Material

The fruit is rounded in shape and the fruit carp is very hard. Fruis are joined with sticks and children use it as wheels and pull as cart or vehicle.

The leaflets are long and easy to tear. Children use the leaflets to make watch and trumpets.

The fruits have also food value and very much nutritive, rich with sugar and protein. The stem is used as firewood and for construction purpose. (Fig. 21 b)

Fig. 21 (a) *Borassus flabellier* L.

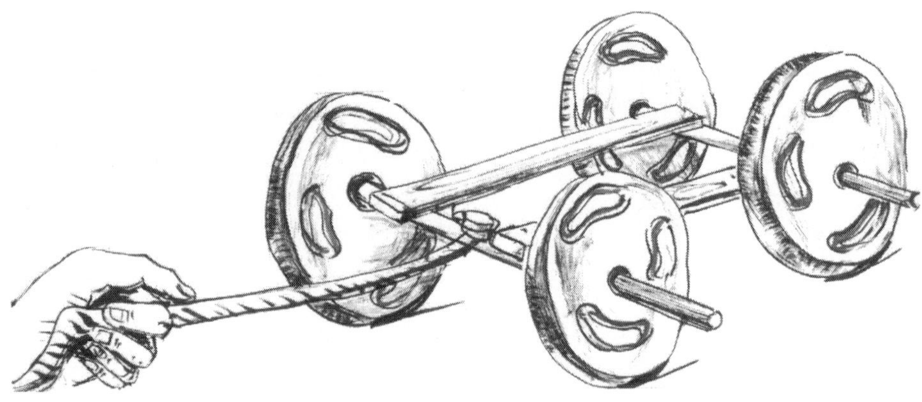

Fig. 21 (b) Children use fruits for making wheel

22

Bryophyllum pinnatum (Lam.) Oken

Common name : English- Miracle leaf

Hindi- Zakhm-haiyal

Odia- Amarpoi

Family : Crassulaceae

Habit & Distribution

The plant is a herb (Fig. 22 a), mostly found in rocky areas and also in plains. It is widely distributed in tropical area.

General Characters

It is a succulent herb, grows upto 1-2m in height. The leaves of the plant are simple, opposite, 3-5 lobed, thick and fleshy with crenate margins. Flowers are pink, or purple in colour. The plant has thick, flexible and short stems. The plant flowers during November to March. Fruiting occurs between January and February.

Part Used as Play Material

The leaf of the plant is fleshy and carries a lot of water. Children use the leaves for cleaning their slates, by folding the leaves and rubbing it on the slates.

Also, the leaves have got medicinal value. The leaf juice is used for the treatment of stomachache, intestinal disorder, one of the home remedies for piles al well. (Fig. 22 b)

Fig. 22 (a) *Bryophyllum pinnatum* (Lam.) Oken

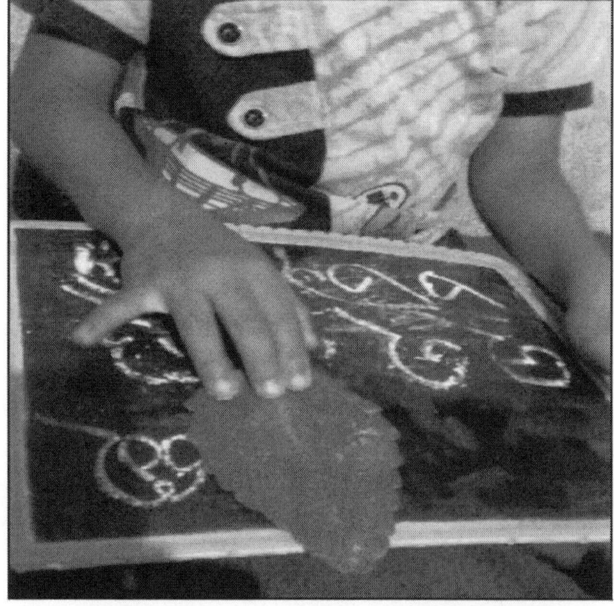

Fig. 22 (b) Leaf are used for cleaning slates

23

Butea monosperma (Lam.) Taub.

Common name : English- Bastard teak

Hindi- Palash

Odia- Palash

Family : Fabaceae

Habit & Distribution

The plant is a tree distributed throughout India, fairly common especially in the drier parts of north-west odisha.

General Characters

It is a small or medium sized dry deciduous tree, reaching to 15 m. in height. It is a slow growing tree. The leaves are pinnate, with about 8-12 cm long petiole and three leaflets, each leaflets of about 10-20 cm long. The flowers are 2.5 cm long, bright orange red, and produced in racemes up to 15 cm long (Fig. 23a). The fruit is a pod. The plant flowers during February and April, and bears fruits during May-July. The plant is popularly known as flame of the forest.

Part Used as Play Material

The leaves are leathery large in size and easy to fold. Children use it for making utensils (plates, bowls) while playing as kitchen set items.

The flowers are bright red in colour and look very attractive. It is used by children for making coloured water, especially in holi festival. The colour made from this flower is harmless and ecofriendly. (Fig. 23 b)

The stem of the plant is used as fuelwood.

Fig. 23(a): *Butea monosperma* (Lam.) Taub

Fig. 23b: Flowers use for making colored water (Especially in holly)

24

Caesalpinia bonduc (L.) Roxb.

Common name : English- Fever nut/Grey Nicker

Hindi- Kat-karanj

Odia- Gila

Family : Caesalpiniaceae

Habit & Distribution

The plant is a herb (Fig. 24 a), commonly found in hedges, thickets, scrub forests etc. and is distributed throughout India.

General Characters

It is a large climber with staright or recurved prickles. The branchlets are hairy and leaflets are 8 pairs with elliptic and lanceolate shape. Flowers are yellow, simple and usually supra-axillary. The pods are broadly oblong with 5-7.5×4 cm in size. It is 1-2 seeded and the seeds are large and round, 1-8cm diameter. The testa is very hard. The flowering time for the plant is August to October and it bears fruits during December to April.

Parts Used as Play Material

The seeds are usually present inside the pod. After maturity, the seeds coming out of the pod are oval shaped, shining and appear like pearls. Children use the seeds as beads for making necklace and bracelets. Further children treat it as the marbles in playing boards (Fig. 24 b).

Fig. 24(a): *Caesalpinia bondue* (L.) Roxb

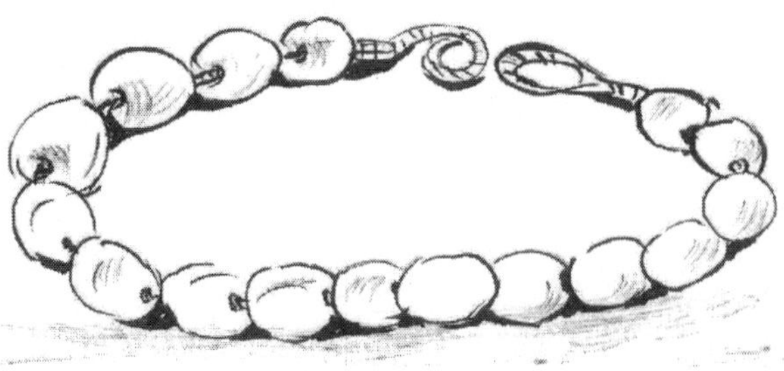

Fig. 24(b): Seeds use for making necklace and bracelets by girls

25

Caesalpinia pulcherrima (L.) Sw.

Common name : English- Peacock flower

Hindi- Guletura

Odia- Godibana

Family : Caesalpiniaceae

Habit & Distribution

The plant is a shrub (Fig. 25a), commonly found in gardens and sometimes in wild. It is cultivated throughout India and probably native to tropical America, now widely cultivated in the tropics.

General Characters

The plant is a branched shrub or small tree that reaches up to 3-4 m in height. The leaves of the plants are bipinnate, 20-40 cm long, bearing 3-10 pairs of pinnae, each with 6-10 pairs of leaflets. The flowers are yellow or red in colour and borne in racemes up to 20 cm long. Petals are four in number, subequal and transversely oblong above the long claw. The plant flowers and fruits during most parts of the year.

Part Used for Play Material

The flowers are red or yellow in colour with spreading shape and looks very attractive. The red petals look like nails of a palm and very much attractive. Children love to use the flower (petals) for making artificial nails (Fig. 25 b).

Fig. 25 (a) *Caesalpinia pulcherrima* (L.) Sw.

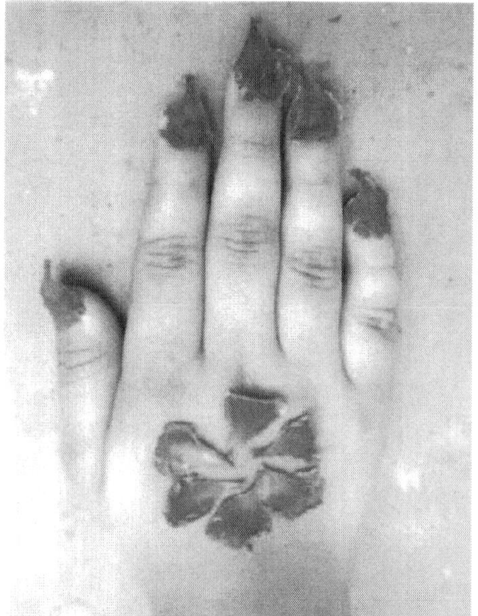

Fig. 25 (b) Children use for making artificial nail

26

Calotropis gigantea R. Br.

Common name : English- Gaint milk weed

Hindi- Akoa

Odia- Arakha

Family : Asclepiadaceae

Habit & Distribution

The plant is a shrub. (Fig. 26 a) The plant is frequently found in waste grounds and roadsides. It is widely distributed throughout India.

General Characters

The plant is a stout-tomentose shrub. Leaves of the plant are 10-16×8-10 cm size, smaller in branchlets with elliptic-ovate to obovate in shape. The fowers are pale purple or greenish white in colour, 3 cm across having 5 pointed petals. The fruit is saccate, 6×3 cm ovoid, inflated and boat shaped. The plant flowers during December and July and bears fruits during February and June.

Part Used as Play Material

The fruit of the plant is oval in shape, light green in colour and having a hard outer covering. The fruits appear like the body of parrot. Children use the fruits for making parrots for fun.

The early flower (bud) of this plant is rounded in shape and filled with air. It makes sound like a cracker when pressed. Children do play with the floral bud by pressing it and often make fun (Fig. 26 b).

The latex of the plant carries medicinal value.

Fig. 26 (a) *Calotropis gigantea* R. Br.

Fig. 26 (b) Fruit is used for making parrot and flower is broken to produce sound play

27

Calotropis procera (Aiton) Dryand

Common name : English- Rubber bush

Hindi- Akada

Odia- Arakha

Family : Asclepiadaceae

Habit & Distribution

It is a shrub (Fig. 27 a), common weed, mostly found in disturbed sites, roadsides, and wastelands and roadsides. It is widely distributed in hot and dry areas of India.

General Characters

The plant is a shrub that grows up to 4 m in height. Leaves of the plant are opposite, gray green, broadky elliptic, large upto 15 cm long and 10 cm broad with a pointed tip. Flowers are waxy white, 5 petals, purple tipped inside and with a central purplish crown. Fruits are gray-green, inflated 8-12 cm long and contain numerous seeds. Flowering and fruiting occur during December and July.

Part Used as Play Material

The fruits of the plant are green and rounded with moderately hard from outside. Children use the fruits as ball for playing different games. (Fig. 27 b)

The stem is long, usually and slightly curved near the base. Children use it as hockey stick.

Fig. 27 (a): *Calotropis procera* (Aiton) Dryand

Fig. 27 (b): Fruit is used as ball

28

Carica papaya L.

Common name : English- Papaya

Hindi- Papaya/Papeeta

Odia- Amruta bhanda

Family : Caricaceae

Habit & Distribution

The plant is a small tree (Fig. 28 a), commonly cultivated throughout India. The plant is widely distributed in tropical and subtropical regions.

General Characters

It is a soft tree. The leaves are large, palmately lobed, long petiole with hollow and swollen base. Sepals and petals are in alternate whorls. The flowers are yellowish-white in colour, and fruits are large and 1 celled. It bears flowers and fruits throughout the year.

Part Used for Play Material

The petiole/leaf stalk of the plant is long with hollow that looks like the shape of a water pipe. Children use the stalk as pipe to suck water, while playing different games. Children also use it to make whistle or trumpet to blow. (Fig. 28 b)

The fruit has got food value and consumed as one of the important vegetables.

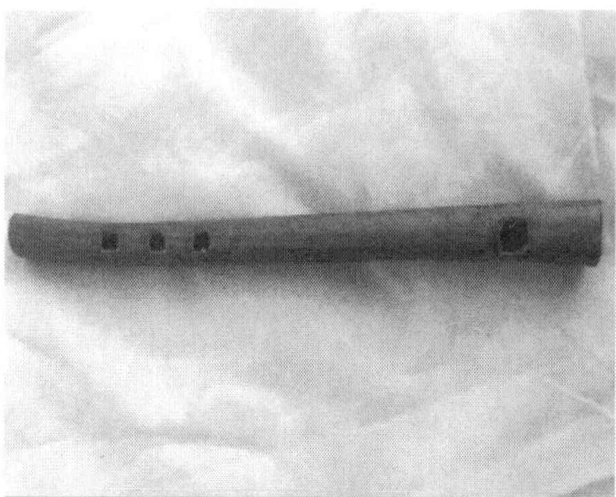

Fig. 28 (a) *Carica papaya* L.

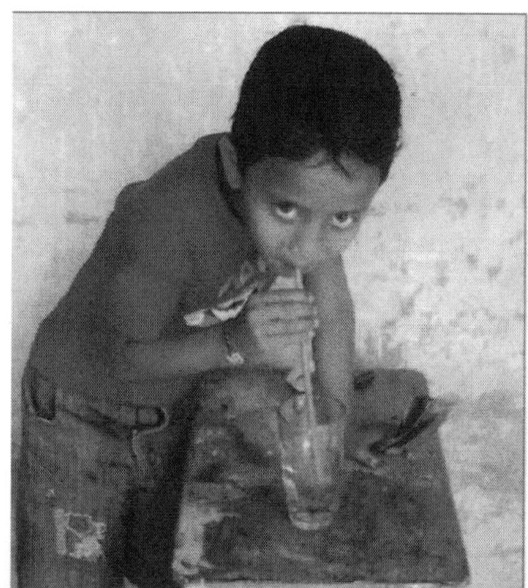

Fig. 28 (b) Leaf stalk is used for making pipe to suck water and to make whistle

29

Cassia fistula L.

Common name : English- Golden shower

Hindi- Sundaraj

Odia- Sunari

Family : Fabaceae

Habit & Distribution

The plant is a medium sized tree (Fig. 29 a), fairly common in forests and often also planted. It is widely distributed throughout India.

General Character

It is a medium sized tree having smooth bark. The leaflets are up to 4-8 pairs, ovate and oblong. Flowers of the plant are light yellow, and at least 30-60 cm long. The fruits are dark brown, long and cylindrical pods. Plants do flower during April and June, and fruiting occurs during March.

Part Used as Play Material

The fruits of the plant are long, cylindrical, light in weight and dark brown in colour. By using two pods as swords, children do fight in games. (Fig. 29 b)

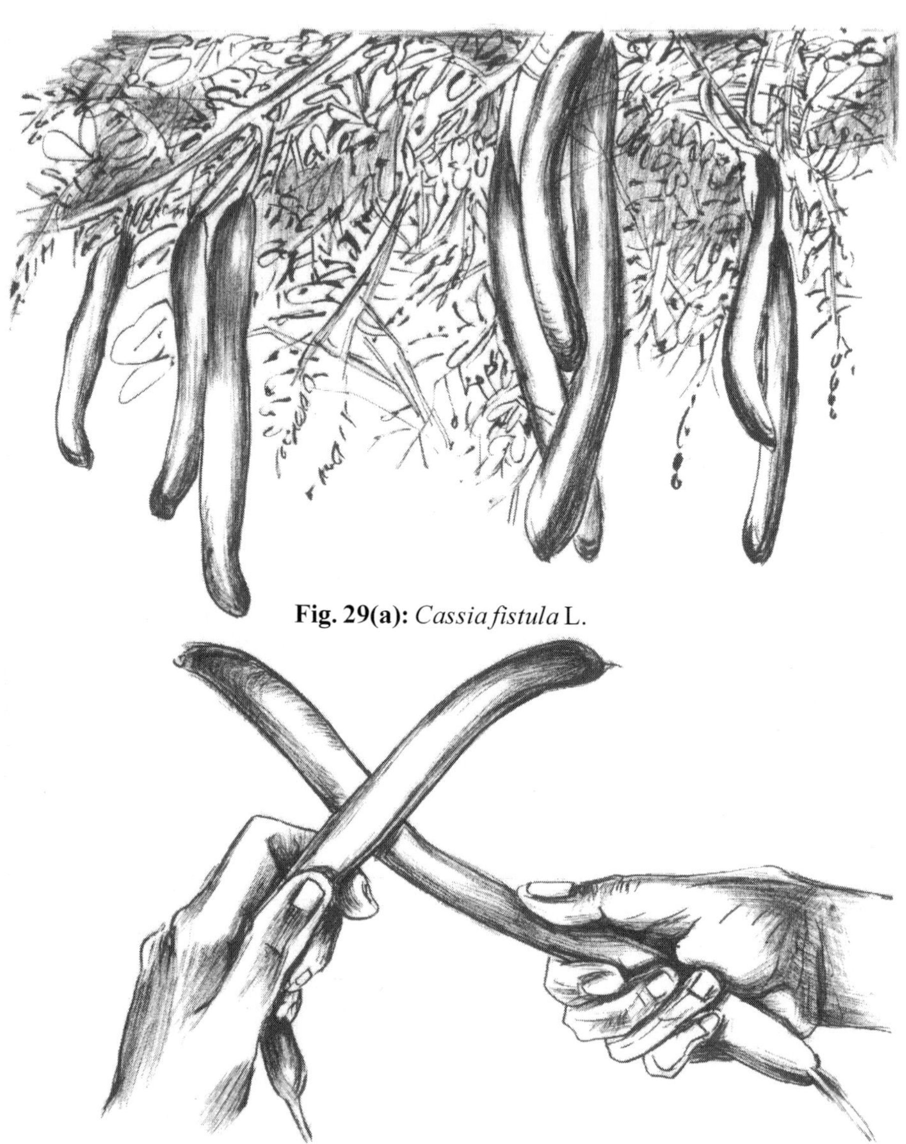

Fig. 29(a): *Cassia fistula* L.

Fig. 29(b): Children use the fruit as swords to play

30

Citrus aurantium L.

Common name : English- Orange

Hindi- Santra

Odia- Kamala

Family : Rutaceae

Habit & Distribution

The plant is a small tree (Fig.30a), commonly cultivated in hilly areas and widely seen in India and other subtropical countries.

General Characters

It is a small tree of moderate size, and grows up to 15-20 ft in height. Thc branches are slender, erect or spreading and spinous. The leaves of the plants are shiny and leathery, oblong to elliptic and grow up to 10 cm long having narrow wings in their petioles. The flowers are white in colour and are seen arranged in clusters of 1-6. Fruits are variable in size and shape, oblate and 7.5 cm in diameter ripen to light or deep yellow in colour. The plant flowers during January to June and bears fruits during November and December.

Part Used as Play Material

The plant body is spinous and spines are seen distributed all over the plant surface. The spines of these plants are about 2-3 inches in size and are pointed just like a needle. Children play with it as needle of a syringe while playing like a doctor. Children put the spines on the arrow head also. (Fig. 30 b)

The epicarp of the orange carries oil glands. Children press the epicarp before the lamp and the light glows spreading outward.

Fig. 30 (a) *Citrus aurantiun* L.

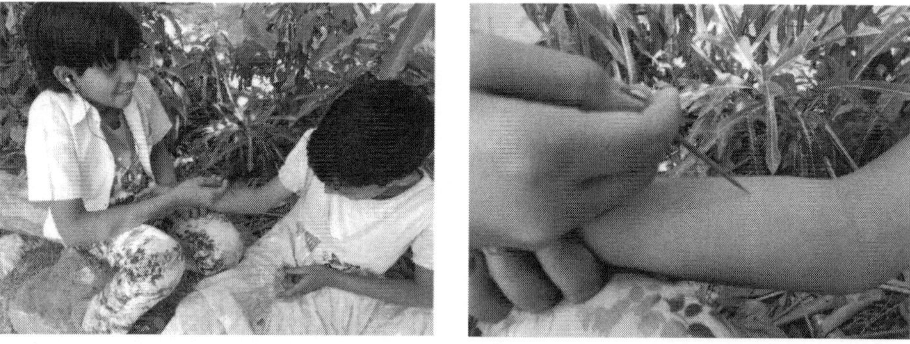

Fig. 30 (b) Throns used as injection while playing doctor's game

31

Clerodendrum infortunatum L.

Common name : English- Hill glory bower

Hindi- Bhant

Odia- Genguti

Family : Verbenaceae

Habit & Distribution

The plant is a shrub (Fig. 31a), common in shady waste places, edges of forests, waysides etc. It is distributed throughout India.

General Characters

It grows up to 1-2 m high, with quadrangular branches covered with silky hairs. Leaves are oval, hairy and oppositely arranged with 10-20 cm long. The base of leaves is heart shaped. Flowers are white, tinged with pink, occur in large panicles. The 5 white petals are tinged pink at base. Four long stamens, 3 cm in size, protrude out of the flower. The plant flowers during January-March and bears fruits during April-July.

Part Use as Play Material

There are 4 white and long stamens seen in the flower. Children usually play with the stamens by doing fight between two stamens to break each other. (Fig. 31 b)

Fig. 31 (a) *Clerodendrum infortunatum* L.

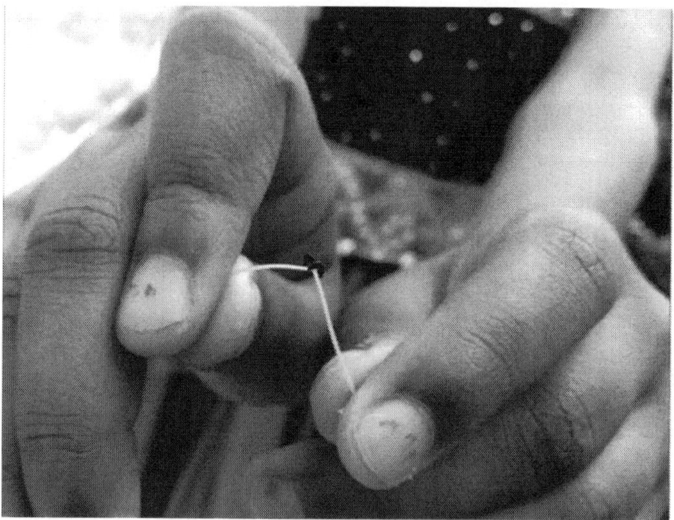

Fig. 31 (b): Stamens used for playing fight game between two stamens to break each other

32

Cocos nucifera L.

Common name : English- Coconut
Hindi- Nariyal
Odia- Nadia

Family : Arecaceae

Habit & Distribution

The plant is a large palm (Fig. 32 a), commonly cultivated, especially in the coastal belts. It is cultivated in the hot and damp regions of India mostly near the sea.

General Characters

It is a large palm reaching up to 30 m in height with pinnate leaves, 4-6 m long and with pinnae 60-90 cm long. The fruit of the plant is a drup and it carries 3 layers; the exocarp, mesocarp and endocarp. The plant has fibrous root system. The tree bears both male and female flowers on the same inflorescence, thus the plant is monoecious. It bears flowers and fruits throughout the year.

Parts Used as Play Material

The shells are woody and hard oval-circular in shape and brown in colour. Children use the shells as utencils while playing. The two halves of the shells are joined together with a thread and treated as telephone by children. Coconut shells are also used for making different craft items as play material like tortoise. The shells are also used as musical instruments (kendara).

The leaves are pinnate and long. So, the leaflets of the plant are used to make a variety of playmaterial like watch, necklace, trumpet, hat, baskets and many other things.

Young coconut fruits are used to make cart by repeated joining the hard veins (khadika) of the leaves. (Fig. 32 b)

Coconut has got food value and it is very nutritive rich with fat. It is also used as holy fruit in different festivals, pujas having religious importance.

Fig. 32 (a) *Cocos nucifera* L.

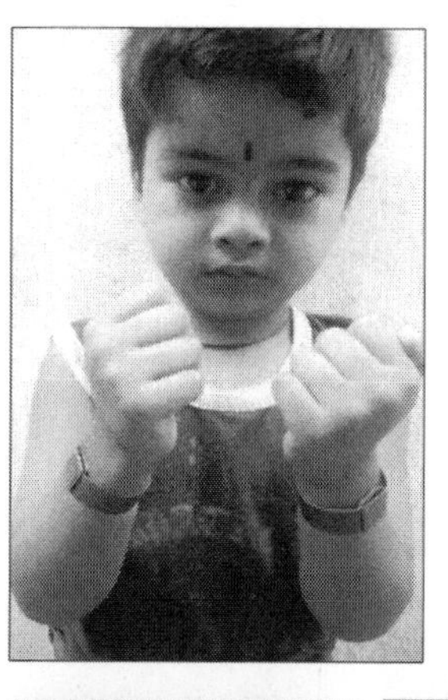
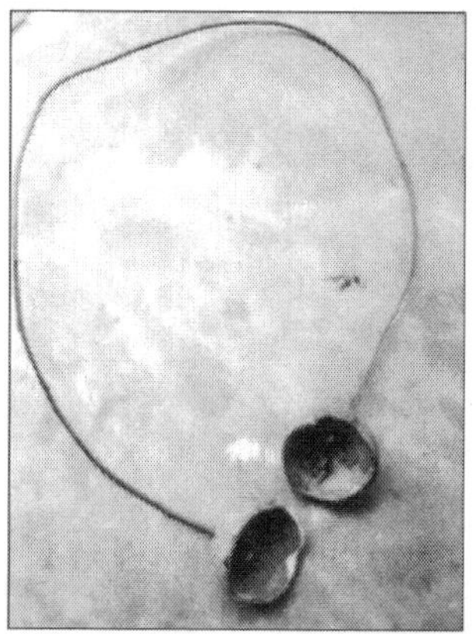
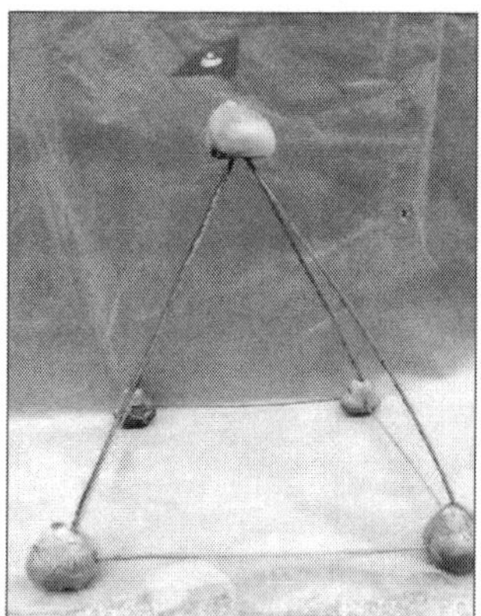
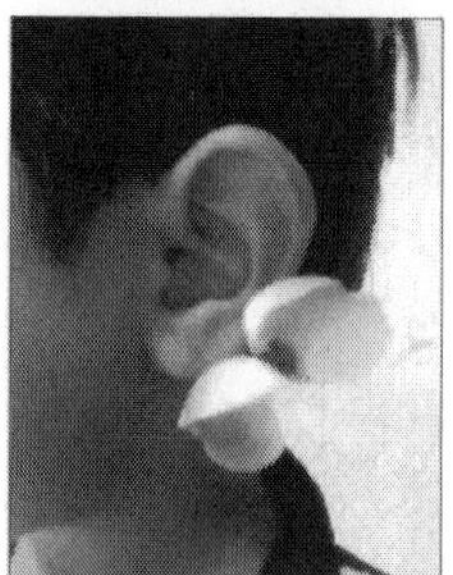

Fig. 32 (b) Shells used as utensils in kitchen set, telephone and leaves are used to make watch, necklace, trumpet etc. Young coconut used for making cart

33

Codiaeum variegatum (L.) Rumph. Ex A. juss

Common name : English- Croton

Hindi- Croton

Odia- Baska

Family : Euphorbiaceae

Habit & Distribution

It is an evergreen shrub (Fig. 33 a) widely planted in gardens and also cultivated. It is distributed throughout tropical region.

General Characters

The plant is a shrub which grows up to 10 ft in height and having large, leathery and shiny leaves. Depending on the cultivar, the leaves may be ovate to linear, entire to deeply lobed and variegated with green, white, purple, orange, yellow, red or pink colours. It bears long clusters of small inconspicuous white or yellow flowers. The plant bears flowers and fruits during dry seasons (March-June).

Part Used as Play Material

The plant has attractive colourful leaves with a combination of green, yellow, red, puple and pink. Because of colour diversity, children use the leaves as ornaments for decoration. In festivals such as Ganesh puja and Saraswati puja the leaves are used for decoration purpose. The leaves are often used for preparing flower bouquets. (Fig. 33 b)

Fig. 33 (a) *Codiaeum variegatum* (L.) Rumph. Ex A. juss

Fig. 33 (b) Children Use the leaves for making flower bookies

34

Coix lacryma-jobi L.

Common name : English- Adlay/Job's tear

Hindi- Gurlu

Odia- Garagada

Family : Poaceae

Habit & Distribution

The plant is a herb (Fig. 34 a), usually found near water bodies, way sides of agricultural lands etc. It is widely distributed throughout the hot and damp parts of India and commonly seen in tropics.

General Characters

The plant is an annual grass rooting at the lower nodes, spongy, glabarous and grows up to 0.9-2 m in height. The leaves are flat, linear, and acuminate with a strong midrib and 10-40 cm long and 205-4 cm in width. The spikes are 6-10 cm long, erect and penduncled. The fruits are hard, bony, ovoid, shining white or nearly black in colour. It bears flowers during September to November and carries fruits during December to February.

Part Used as Play Material

The fruits of the plant are shining white in colour with ovoid in shape,that look like small eggs. Children use the fruits as egg while playing kitchen games. Further by mounting in thread, children do prepare necklace. (Fig. 34 b)

Fig. 34 (a) *Coix lacryma-jobi* L.

Fig. 34 (b) Seeds are used as egge while playing kitchen game

35

Colocasia esculenta (L.) Schott

Common name : English- Elephant ear taro

Hindi- Kachu/Kachlu

Odia- Saru

Family : Araceae

Habit & Distribution

It is a herb (Fig. 35 a), commonly found in damp, shady places and also cultivated. The plant is widely distributed throughout the hotter parts of India.

General Characters

The plant is a tuberous bulb plant growing about 3-5 ft in height. The plant has large leaves. The leaves are heart shapes, 2-3 ft long with a long stalk of about 3 ft height. The flowers are yellowish-white spathes and spadixes are frequently produced. It bears flowers and fruits during June-November.

Part Used as Play Material

The leaves are very large in size like the ear of an elephant. Therefore, children use the leaves to capture dragonfly, by holding two leave one on each hand. The large size and surface area of the leaves help the children to easily catch the dragon flies. (Fig. 35 b) Further water does not stick to the surface of the leaves and children play with it like pearl.

Fig. 35 (a) *Colocasia esculenta* (L.) Schott

Fig. 35 (b) Leaves are used to capture dragonfly and play water bubble game

36

Corchorus capsularis L.

Common name : English- White jute

Hindi- Titapat

Odia- Jhota

Family : Tiliaceae

Habit & Distribution

It is a herb (Fig. 36 a), commonly cultivated throughout India for fiber purpose.

General Charcaters

The plant is an erect, annual herb reaching up to 1-2 m height. The stems are usually purplish white. Leaves are ovate-lanceolate, 5-12 cm long, pointed at the tip and rounded at the base. The flowers are borne in small groups in the axils of the leaves about 4 mm long. The capsules are round to ovoid and about 1 cm in diameter. The plant flowers and bears fruits during July-November.

Part Used as Play Material

The fibre is extracted from the cut stems after retting in water. The stem (stem) is dried, that beccomes very light and whitish in colour. As it is whitish and yellow, children often treat the stem pieces as cigarette (Fig. 36 b). Of course this does not carry any nicotine. Further they also often use it (stem) as pipe/ straw to suck water/juice.

Fig. 36 (a) *Corchorus capsularis* L.

Fig. 36(b) Stem is used for making cigarette, straw/pipe as play material

37

Cordia dichotoma G. Forst.

Common name : English- Indian cherry

Hindi- Lasora

Odia- Ambota/Gualikoli

Family : Boraginaceae

Habit & Distribution

The plant is a small tree (Fig. 37a), found frequently in mixed forests and distributed throughout India.

General Characters

The plant is a small tree with ususaly drooping branches and spreading crown. The stem bark is grayish brown, smooth or longitudinally wrinkled. Leaves are orbicular, elliptic-oblong, entire with 6-12.5 cm long. Flowers are white, pinkish white, fragrant, short stalked. The fruit is a yellow shining globose or ovoid drupe seated in a saucer like enlarged calyx and it turns black on ripening and the pulp gets viscoid. The plant flowers during March-April and bears fruits during July-September.

Part Used as Play Material

There is a jelly and sticky like juice present inside the fruit of the plant. This sticky juice is collected by rupturing the fruit and used as gum by the children to prepare different play material like kites, paper boats, packets, book covers. (Fig. 37 b)

Fig. 37 (a) *Cordia dichotoma* G. Forst.

Fig. 37 (b) Fruit extracts are used as gum to prepare kites

38

Crotalaria pallida Aiton.

Common name : English- Rattle pods

Hindi- San

Odia- Junujhunuka

Family : Fabaceae

Habit & Distribution:

It is a shrub (Fig. 38 a), common in waste places, roadsides and river banks. It is widely distributed in India.

General Characters

The plant is a perennial shrub, with stems ascending or erect, 1-2 m tall. Leaves are trifoliate, leaflets elliptic, ovate, 2.8-9.5 cm long and 1.2-4.5 cm wide. Flowers are yellow, 1-1.5 cm long, somewhat long inflorescence about 15-40 cm long. Pods brown at maturity, 3.5-3.8 cm long. The plant flowers and bears fruits during April.

Part Used as Play Material

Pods are soft and the inner side is filled with air. Children use these pods as cracker as it creates sound like cracker when it is broken up. (Fig. 38 b)

Fig. 38 (a) *Crotalaria pallida* Aiton.

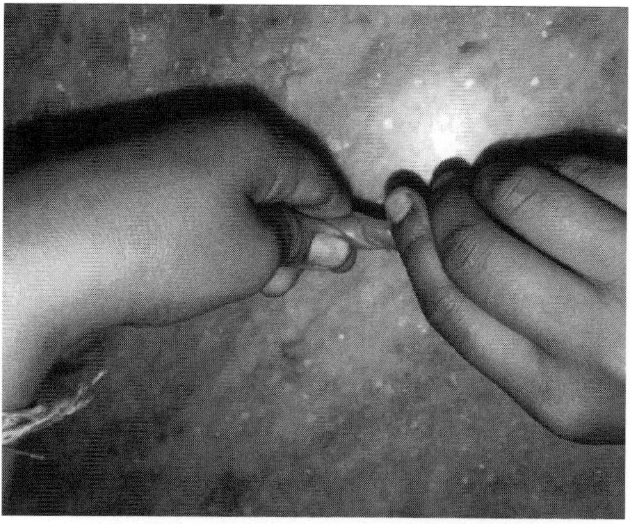

Fig. 38 (b) Children use the fruits as crackers

39

Cucumis sativus L. var. *hardwickii* (Royle) Gabaev

Common name : English- Wild cucumber
Hindi- Khira
Odia- Pita kakudi

Family : Cucurbitaceae

Habit & Distribution

It is a herbaceous climber (Fig. 39a) apparently wild in open waste lands and distributed all over India.

General Characters

The plant is a herbaceous climber with simple tendrils. Leaves are entire or shallowly 5 lobed with 12 cm across and 15 cm petiole. Flowers are yellow in colour, 1.7-2.5cm across. The fruits are oblong, yellowish-green, glabrous and covered with rigid hairs or soft spines. Flowering and fruiting occur during September-November.

Part Used as Play Material

The fruits are oval shaped. Children use the fruits to make toys of different shapes like horse, chicks etc as their play material by inserting sticks in to the fruits. (Fig. 39 b).

Fig. 39 (a) *Cucumis sativus* L. var. *hardwickii* (Royle) Gabaev

Fig. 39 (b) Fuits are used for making toys like Chickens, horses etc.

40

Cuscuta reflexa Roxb.

Common name : English- Giant dodder

Hindi- Amar bel

Odia- Nirmuli

Family : Cuscutaceae

Habit & Distribution

It is a parasitic plant (Fig. 40 a) that usually grows in a profilic manner over other plants as support. It is widely distributed throughout India.

General Characters

The plant is a creeper with yellowish-green or yellow stem and often in dense mass. The flowers of the plant are white or pinkish and almost 6-7.5 mm long. The corolla of the flower is tubular with short reflected lobes. The fruits are fleshy and globose with conical in shape. It bears flowers during October to December and bears fruits during December-Jnauary.

Part Used as Play Material: Stem (vines)

The plant stem is yellowish in colour and look like long thread. Children collect the stem and cut into pieces and make noodles in kitchen games. Further Chlidren hang the threads on hairs as golden hairs. (Fig. 40 b)

Fig. 40 (a) *Cuscuta reflexa* Roxb.

Fig. 40 (b) Vines are used as noodles while playing kitchen game

41

Cynodon dactylon (L.) Pers.

Common name : English- Bermuda grass

Hindi- Dub

Odia- Duba ghasa

Family : Poaceae

Habit & Distribution

The plant is fairly common in lawns, roadsides, cultivated lands, damp grounds, wastelands etc. It is widely distributed throughout India. (Fig. 41 a)

General Characters

It is a variable perennial creeper with hairy leaves, 3-7 spikes (rarely 2) together at the top of the stem. The stem is erect and often grows 1-30 cm in height. It has a deep root system that grows to over 2 m deep in to the soil. It flowers and bears fruits throughout the year.

Part Used as Play Material

The shoot of the plants are erect and flattened with hairy leaves. Children prepare brooms by collecting a number of shoots and making bundle, binding it at the bottom with a thread. They use it for sweeping while playing games of different types. (Fig. 41 b)

The plant has socio-cultural importance and used in many social functions. The plant is having medicinal impotance and used to stop bleeding.

Fig. 41 (a) *Cynodon dactylon* (L.) Pers.

Fig. 41 (b) Shoot are used for sweeping while playing

42

Datura metel L.

Common name : English- Thron apple

Hindi- Dhatura

Odia- Duddura

Family : Solanaceae

Habit & Distribution

The plant is a shrub (Fig. 42 a), commonly found in road sides, waste lands, damp grounds and also in forests. It is widely distributed throughout India and is of cosmopolitan nature.

General Characters

It is an annual shrub growing up to a height of 8 ft. The stem is green or purple in colour and leaves are ovate and irregularly lobed about 5-12.5 cm long. Flowers of the plant are white or purple in colour and 6.2-12.5 cm in size. The fruit is an egg shaped spiny capsule almost 5 cm in diameter. The plant flowers and bears fruits in the month of November.

Part Used as Play Material

The fruit is rounded in shape with having small spines all over it. It just looks like a play ball. Children use it as a ball for playing catch and throw games. As the fruit carries spines on its surface, children use to pinch each other by this fruit for the shake of fun. (Fig. 42 b)

Fig. 42 (a) *Datura metel* L.

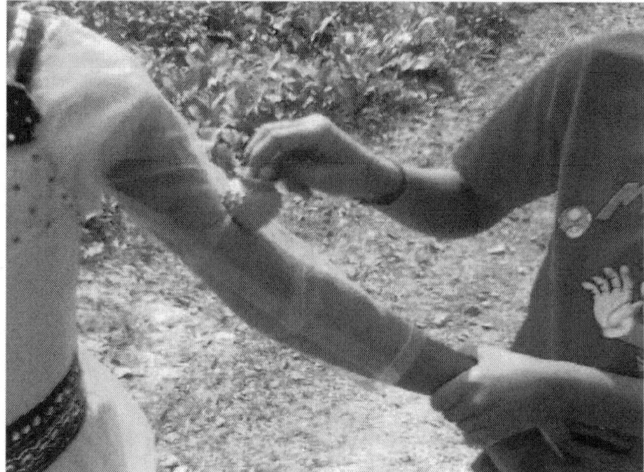

Fig. 42 (b) Children use the fruits for punching one another in order to make fun.

43

Dalbergia sissoo DC.

Common name : English- Rose wood

Hindi- Shisham

Odia- Sissoo

Family : Fabaceae

Habit & Distribution

It is a tree (Fig. 43 a), commonly planted along roadsides. The plant is widely distributed throughout India.

General Characters

The plant is a medium sized tree with thick, gery or pale brown bark. It reaches up to a maximum of 25m in height and 2-3 m in diameter. The leaves are leathery, alternate and about 15 cm long. The flowers are yellowish in colour and 1.5 cm in long. The pods are oblong, flat, thin and 2.5-7.5 cm long with 3-4 seeds in it. The plant flowers during March-June and bears fruits during November-February.

Part Used as Play Material

The stem of the plant is very strong, woody and erect. Children use the wood and stem to make bats for playing cricket. It is very plain, strong and long lasting. The wood is also used for making bat (in gilli-danda) as it is extremely durable. This wood is very resistant to termites. (Fig. 43 b)

Fig. 43 (a) *Dalbergia Sisso* DC.

Fig. 43 (b) Stem is used for making cricket bats

44

Delonix regia (Hook.) Raf.

Common name : English - Gold mohar

Hindi - Gul mohar

Odia - Krusnachuda

Family : Fabaceae

Habit & Distribution

The plant is a tree (Fig. 44 a), commonly planted in gardens and avenues. It is widely planted throughout India as a street tree.

General Characters

The plant reaches up to 30-40 ft in height. Leaves are feathcry and grow up to 60 cm long with 11-18 pinnae pairs and many leaflets. The flowers are orange or red in color and more than 10 cm across and appear in corymbs along and at the ends of branches. Pods are 30-60 cm long, green and flaccid when young and turn dark-brown later. The plant bears flowers in the month April-June.

Part Used as Play Material

The pods are dark brown in colour and long in size. It is very light. Children use it as swords.

The stamens are nearly equal or shorter than the petals and can easily be removed from the flower. Chlidren use the stamens for playing, by making fight between two stamens. (Fig. 44 b)

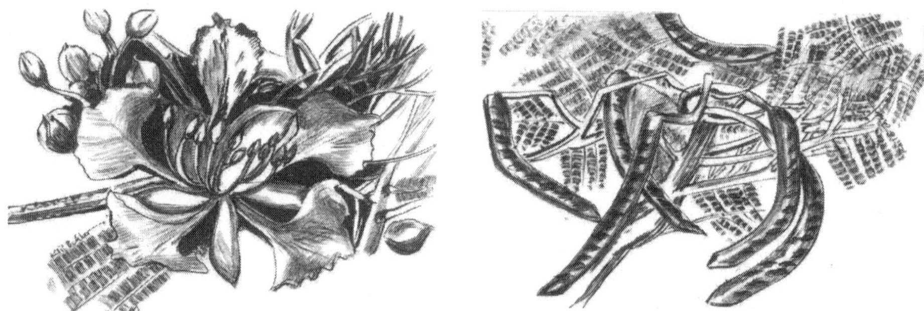

Fig. 41 (a) *Delonix regia* (Hook) Raf.

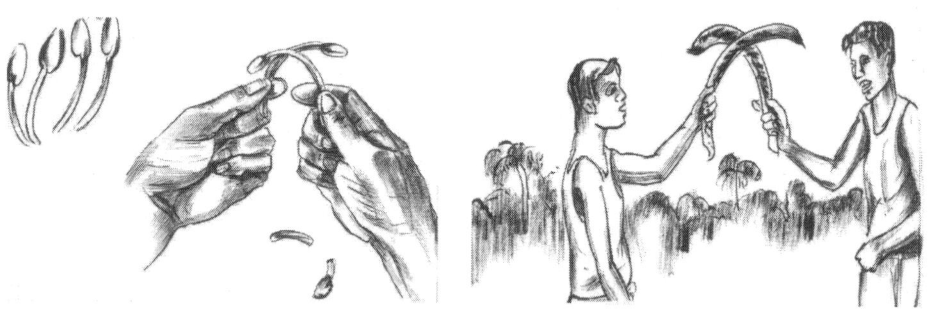

Fig. 43 (b) Stamens used for playing by doing flight between two stamens to break each other and the pods as Swords to play

45

Dillenia indica L.

Common name : English- Elephant apple

Hindi- Cwhalta

Odia- Awoo

Family : Dilleniaceae

Habit & Distribution:

It is a tree (Fig. 45 a) which is frequently planted and often grown in gardens. The plant is widely distributed in evergreen forests throughout India.

General Characters

The plant is a moderate sized evergreen tree growing up to 15 in tall. The leaves are deep green in colour, oblong to lanceolate and 15-36 cm in long. The flowers are white, large, up to 5 inches across/ the fruit is rounded, 5-12 cm in diameter aggregate of 15 carpels and each carpel containing 5 seeds embedded in the pulp. The plant flowers during May- June and bears fruits in September-February.

Part Used as Play Material

The fruits are rounded in shape and high weight. Children use this fruit in games such as short foot throw. The fruits are edible and are very nutritive. The fruits are eaten by local people. (Fig.45 b)

Fig. 45 (a) *Dillenia indica* L.

Fig. 45 (b) Fruits are used as short foot for throwing

46

Drypetes roxburghii (Wall.) Hurus.

Common name : English- Lucky bean tree

Hindi- Putija/Putranjiva

Odia- Poitundia/Poichandia

Family : Euphorbiaceae

Habit & Distribution

It is a tree (Fig. 46 a), growing in hills, in wild often planted and self shown. It is distributed throughout tropical India.

General Characters

The plant is a moderate sized evergreen tree growing up to 12 meter in height. Leaves are simple, alternately arranged, dark green and shiny in colour. There are two types of flowers with short stalk, in rounded axillary clusters; females are green and 4 mm across. Fruits are ellipsoid or rounded drupes, white velvetly, seed normally one, stone pointed and very hard. The plant flowers during March-April and bears fruits during January-March.

Part Used as Play Material

The fruits of the plant are rounded in shape with a pointed end, which appear like a 'Natu'. Children use the fruits as natu, as their play material and rotate on the floor. (Fig. 46 b)

Fig. 46 (a) *Drypetes roxburghii* (Wall.) Hurus

Fig. 46 (b) Fruit is used to make 'Natv' for playing

47

Erythrina variegata L.

Common name : English- Indian Coral tree

Hindi- Pangara

Odia- Paladhua

Family : Fabaceae

Habit & Distribution

It is a small tree (Fig. 47 a), commonly planted and also found in wild in coastal regions of India.

General Characters

The plant is a prickley tree with smooth and greenish bark. Leaflets are rhomboid-ovate, 5-15cm×5.5-17 cm and acute or acuminate in shape. The flowers are deep red in colour, scarlet, 7 cm long, calyx split and spathaceous with very oblique mouth. Pods are many, subcylindric, narrowed with 15-25 cm long. It is 6-12 seeded, seeds are 1.5-1.8 cm in size with dark purple in colour. The plant flowers during March-April and bears fruits during April-July.

Part Use as Play Material

The flowers of the plant are deep red in colour and look very attractive. Children use it for different decoration purposes. They also use it for colouring their hands. (Fig. 47 b)

Fig. 47 (a) *Erythrina variegata* L

Fig. 47 (b) Flower use for colouration of palm and decoration

48

Eucalyptus globulus Labill.

Common name : English- Blue gum
Hindi- Nilgiri
Odia- Eukalyptus/Nilgiri

Family : Myrtaceae

Habit & Distribution

It is a tree (Fig. 48 a). The plant commonly planted in gardens and elsewhere and also widely cultivated as a plantation crop by forest department. It is commonly seen in most parts of India.

General Characters

The plant is a large, evergreen tree and reaches up to the height of 60 m. The leaves are commonly lanceolate, long-petiole, apperantly alternate and waxy green. The flowers have numerous fluffy stamens which may be white, cream, yellow, pink or red in colour. In the bud, the stamens are enclosed in a cap known as operculum, composed of fused sepals and petals. The plant bears flowers and fruits during January-August.

Part Used as Play Material

In the bud, the stamens are enclosed in a cap known as an operculum which is composed of the fused sepals and petals. The operculum is initiated as the tooth (canines) and the dried ovary is used as the top by the children. (Fig. 48 b)

The plant has lots of economic value. The wood is used for pulp making for paper industry.

Fig. 48 (a) *Eucalyptus globulus* Labill.

Fig. 48 (a) Operculum of stamen is used as toot (Canines) and dried ovary used as top

49

Ficus benghalensis L.

Common name : English- Banyan tree

Hindi- Barh

Odia- Baro

Family : Moraceae

Habit & Distribution

It is a tree (Fig. 49 a), fairly common, seen, in wild and sometimes planted. The plant is widely distributed throughout the plains of India.

General Characters

The tree is very large, growing up to 30 m tall, with many aerial roots that develop in to new trunks, so that the tree goes on spreading laterally, indefinitely. A single tree can cover a very wide area. The leaves are leathery, entire, ovate or elliptic, which grow up to 20-30 cm long with prominent lateral viens. The fruits are 1 to 2 cm in diameter, rounded in shape, without stalks arranged in pairs at leaf axils and it changes to bright red colour when ripens. The plant flowers round the year and bears fruits during April-June.

Part Used as Play Material

Leaves: The leaves are large, leathery and ovate in shape. So, it is easy to fold to make different shapes. Children use it for making utensils, umbrellas of different shapes as their play materials.

Prop roots: The prop roots are very long, spread all over the area covered by the plant. Modified roots do originate from the branches of the mother plant and stand in hanging. Children use these roots in order swing from one point to other because of its strength and flexibility (Fig.49 b).

The plant is used as a fire wood. It provides shades during summer to men and other animals. The leaf is used to teach venation in classroom leading. The prop root used to teach root modification.

Fig. 49 (a) *Ficus benghalensis* L.

Fig. 49 (b) Leaves are used to make utensils and umbrellas and prop roots are used as the swings.

50

Ficus racemosa L.

Common name : English- Cluster fig tree

Hindi- Goolar

Odia- Dimiri

Family : Moraceae

Habit & Distribution

The plant is a tree (Fig. 50 a), fairly common in valleys and village surroundings. It is widely distributed throughout India.

General Characters

It is a large sized tree with few short areal roots. Leaves are ovate or elliptic with 7-20×3.5-9 cm in size. The most distinctive aspect of this tree is the red, furry figs in short clusters, which grow directly out of the trunk of tree. The figs do carry hundred of flowers in it. The plant flowers during the month March-June and bears fruits during July-September.

Part Used as Play Material

The fruits are small, rounded, equal size and shape. Children use the fruits as wheels for making Chariots by interconnecting the fruits by coconut leaf veins (khadika). They pull the chariot by rope. (Fig. 50 b)

The plant also used for fire wood purpose.

Fig. 50 (a) *Ficus racemosa* L.

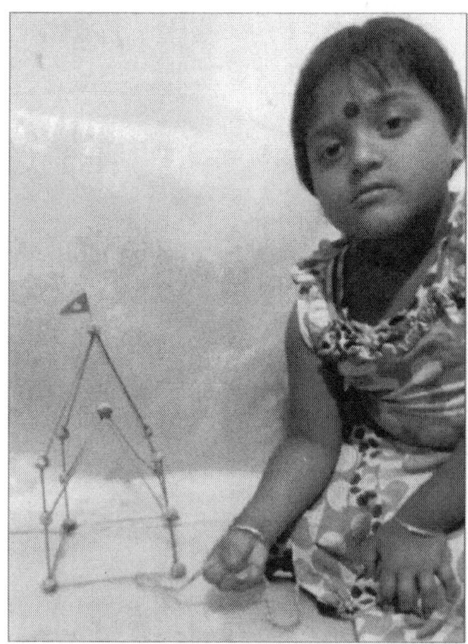

Fig. 50 (b) Fruits are used for making choriots

51

Ficus religiosa L.

Common name : English- Pipal tree

Hindi- Peepal

Odia- Aswatha

Family : Moraceae

Habit & Distribution

The plant is a large tree (Fig. 51 a), found in common, wild or human setteled localities. The plant is widely distributed throughout India.

General Characters

It is a large tree with grey bark and numerous roots which have fused with stem. Leaves are leathery, 4-8 inches long by 3-5 inches width with tailed at the tip and heart shaped at the base. Fruit is a fig which is axilary, sessile, 1-1.5 cm diameter and turns into dark purple when ripens. The plant flowers during June to October.

Part Used as Play Material

The leaves of these plants are leathery and broad. Children use it to make whistles by rolling the leaves as their play material. As the leaves are broadly ovate in shape, it is used for decoration purposes by the children in different festivals.

The children use the leaves to see the nature of venation and also for making new year greeting cards.

In villages, the people used to put swings on its branches (Fig.51 b). The plant has also religious importance and the people in Hindu culture worship it. Further it provides shade and shelter to animals like monkeys, birds and insects.

Fig. 51 (a) *Ficus religiosa* L.

Fig. 51 (b) Leaves are used to make whistles, decoration, greeting cards, and swings on branches

52

Impatiens balsamina L.

Common name : English- Garden Balsam

Hindi- Gul-mendi

Odia- Haragaura

Family : Balsaminaceae

Habit & Distribution

It is a herb (Fig. 52 a) commonly cultivated in gardens and also self-sown in waste lands, forest edges, roadsides. It is distributed throughout India.

General Characters

The plant is an erect annual herb, growing upto 30-80 cm in height with stilt roots from the lower nodes. The leaves are alternate, lanceolate or oblong-lanceolate and 2.5-9 cm long and 1-2.5 cm broad. Flowers are pink, red and white in colour with 2.5-5 cm in diameter. Flower stalks are up to 1-5 cm long. Capsules are ovoid ellipsoid and 1.5-2.5 cm long. The plant flowers and bears fruits during August-December.

Part Used as Play Material

The flowers of this plant are seen in several colours like red, pink, purple, white etc. which look very attractive. Children usually use these flowers for coloration of palms by chopping it and putting its paste. Besides this the flowers are also used for different decoration purposes during festivals and pujas. (Fig. 52 b)

Fig. 52 (a) *Impatiens balsamina* L.

Fig. 52 (b) Children use the flowers for colouration of palms and other decoration purposes

53

Ipomoea hederifolia L.

Common name : English- Scarlet morning glory
Hindi- Lalpungli
Odia- Panikod

Family : Convolvulaceae

Habit & Distribution

The plant is (Fig. 53 a) a herbaceous climber, common in open forests and open places over hedges, bushes, fairly naturalized. It is naturalized throughout India.

General Characters

It is a twining, smooth to hairy annual plant. Leaves are ovate or broadly ovate, heart shaped having bases with pointed tips. The flowers are red in colour with a longer stalk than the leaves. Each flower stalk may bear a single flower or may have several flowers. The fruit is a round capsule up to 8 mm in diameter containing few seeds. Flowering and fruiting occurs during September-January.

Part Used as Play Material

The flowers are red in colour. Children use it to decorate their forehead and also use to make different ornaments for making fun. (Fig. 53 b)

Fig. 53 (a) *Ipomoea hederifolia* L.

Fig. 53 (b) Flowers used for decoration of Forehead, also used as Ornaments

54

Ipomoea parasitica (Kunth) G. Don

Common name : English- Yellow-Throated Morning Glory

Hindi- Neelkalme

Odia- Kaladana

Family : Convolvulaceae

Habit & Distribution

The plant is an ornamental climber (Fig. 54a), common in waste places and road sides. It is widely distributed throughout the tropical countries.

General Characters

It is a climber. Stems trail for several metres and climb over low bushes, reddish, muriculate, hairless or velvety. The Leaves are ovate, 5-15 × 5-15 cm, blunt or long-pointed at the tip and heart-shaped. Flowers are funnel-shaped, blue with purplish tinge, 3.5-4.5 cm long, with a yellow center and borne in several-flowered cymes. The cyme is carried on a stalk 5-25 cm long. Capsule is ovoid to globose, hairless. Seeds are trigonous, brownish, hairless or very shortly velvet-hairy.

Part Used as Play Material

The leaves are heart shaped. Children use the leaves for different decoration and ornamental purposes by folding the petioles of leaves. (Fig. 54 b)

Fig. 54 (a) *Ipomoea parasitica* (Kunth) G. Don

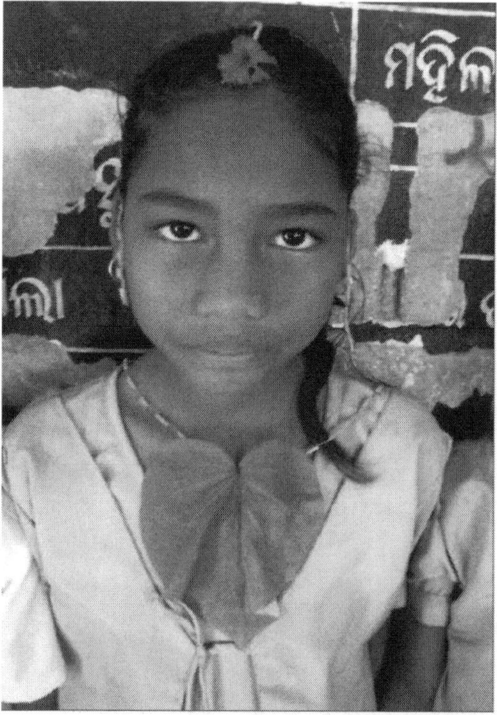

Fig. 54 (b) Leaves are used for making different ornaments for playing

55

Jasminum arborescens Roxb.

Common name : English- Royal jasmin

Hindi- Chameli

Odia- Niali/Banamali

Family : Oleaceae

Habit & Distribution

It is a shrub (Fig. 55 a), commonly found in forests and distributed in Central and Southern India.

General Characters

The plant is nearly erect, not twining or climbing. Branchlets are hairy. Leaves are simple, opposite, variable in size up to 18 cm long and 10 cm broad with long pointed end. Flowers are white, fragrant, three forked clusters with flowers stalk of 5 mm long. Berry is mostly simple, 1-1.5 cm long, ellipsoid and black when ripe. The plant flowers during March-May and bears fruits during June-July.

Part Used as Play Material

The flower is white with an attractive fragrance. Children (especially girls) use it to decorate their hairs (tribal girls mostly use it in chait parab) (Fig.55 b).

Fig. 55 (a) *Jasminum arborescens* Rorb.

Fig. 55 (b) Flower is used for decoration of hair

56

Jatropha gossypiifolia L.

Common name : English- Bellyache bush

Hindi- Ratanjoti

Odia- Baigoba/Baigewa

Family : Euphorbiaceae

Habit & Distribution

It is a shrub (Fig. 56 a), very commonly found in waste ground, village hedges, roadsides and scrub forests. The plant is cultivated and naturalized in several parts of India.

General Characters

The plant is a small dark colored glandular hairy shrub, 0.8-1.8 m in height. The leaves are 6-12 cm in diameter, palmatly 3-5 lobed, entire with 3-5 cm long petiole. Flowers are red with yellow centre. The petioles and stipules have glandular hairs. It flowers and bears fruits during July-October.

Part Used as Play Paterial

The petioles are smooth and long about 5 cm in length. The sticky and foamy liquid/sap present at the petiole and it oozes out when the stem is broken. Children prepare air bubbles by breaking the petiole from the middle part and blowing air into the foamy liquid through a loop from their mouth. It is a well practiced activity to make balloons from jatropha sap by children for making fun. The seeds are used for playing king-queen games (Fig.56 b).

The plant has got economic value. The oil extracted from the seeds is used as bio-diesel.

Fig. 56 (a) ***Jatropha gossypiifolia*** L.

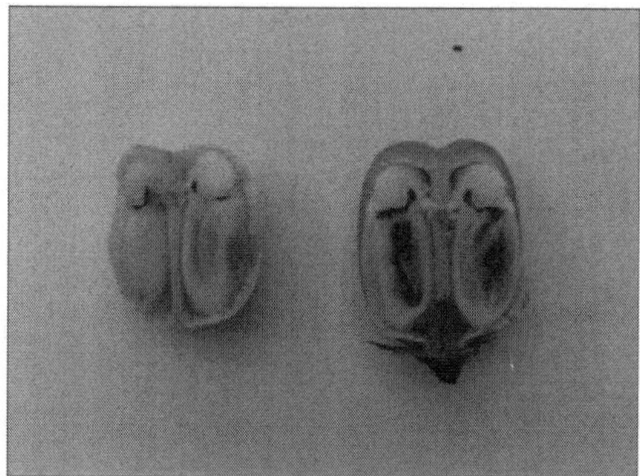

Fig. 56 (b) Children make foam bubbles from sap by breaking the petiole, and Seeds are used to play king-queen games

57

Jatropha integerrima Jacq.

Common name : English- Peregrina/Spicy jatropha

Family : Euphorbiaceae

Habit & Distribution

The plant is a shrub, usually planted in gardens.

General Characters

It is a small erect and branched shrub (Fig. 57 a). The leaves are entire, elliptic-oval, may be fiddle shaped with 7 inches long. The flowers are red and pink and about 2.5 cm across and borne in many flowered clusters at branch end. Fruit is rounded-trigonous. The plant bears flowers and fruits in most parts of the year.

Part Used as Play Material

The stem wood of the plant is soft and durable. Children use it to make baskets and other craft items (Fig.57 b).

Flowers are deep red to pink in colour and look very attractive. Children use the flowers for several ornamental and decoration purposes and for playing different games and fun.

Fig. 57 (a) *Jatropha integerrima* Jacq.

Fig. 57 (b) Stems used for making baskets and flower for Ornamental purpose

58

Lagenaria siceraria (Molina) Standl.

Common name : Odia- Laoo

English-Calbash

Hindi- Lauki

Family : Cucurbitaceae

Habit & Distribution

The plant is a climber (Fig. 58 a), commonly cultivated and distributed throughout India.

General Characters

It is an annual vigorous herb, running or climbing vine with large leaves and with lush appearance. Leaves are simple, 17-20 cm diameter, short and softly hairy with kidney or heart shaped outlines. Flowers are white, up to 4 inches in diameter with spreading petals. Fruits are of various shapes, in different varieties, like bottole or dumbbell shaped, rounded or elongated. The plant flowers during July-January and bears fruits during cold season.

Part Used as Play Material

The epicarp/outer covering of the fruit is very hard and plain from the outside. After consuming the edible part of the fruit, the epicarp of the fruit is used by the children for making play materials like utensils and musical instruments (Lau Duduki) because, it is oval in shape and durable (Fig.58 b).

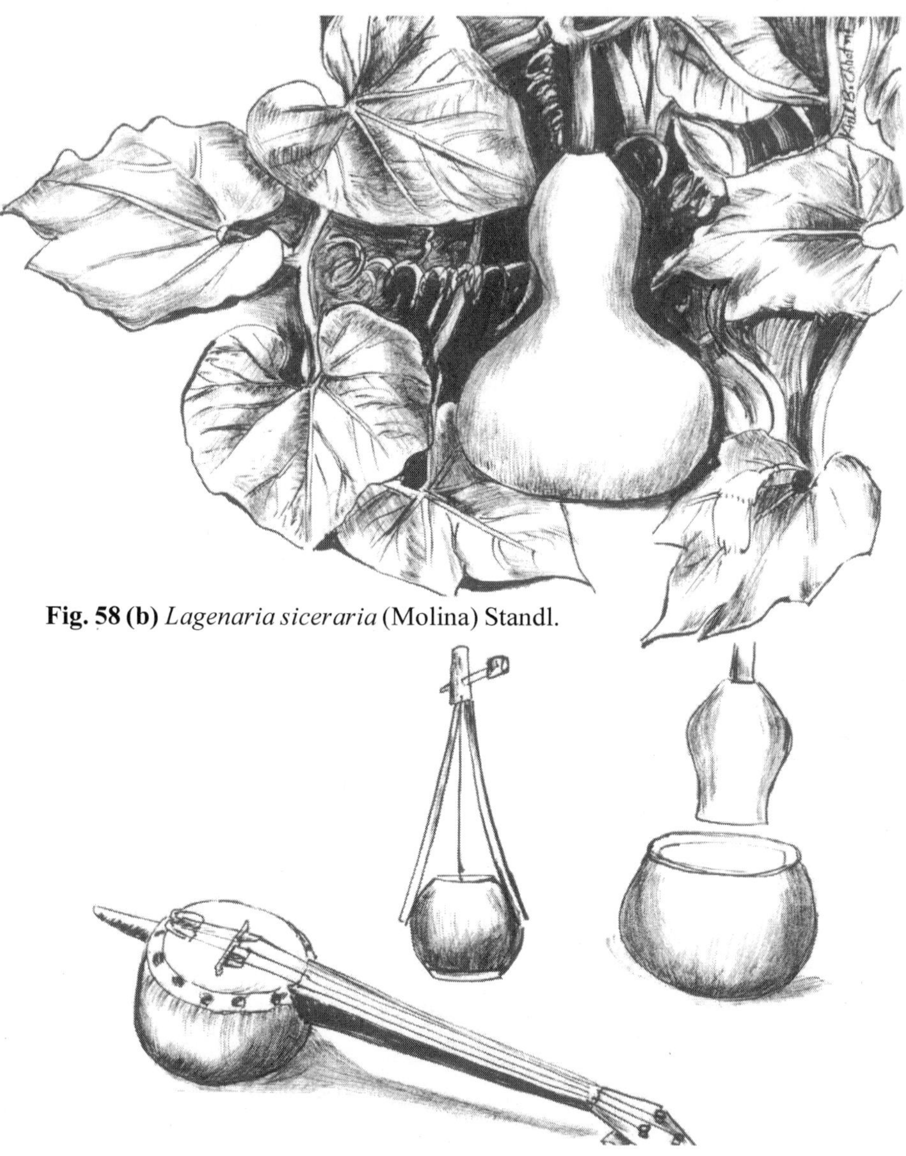

Fig. 58 (b) *Lagenaria siceraria* (Molina) Standl.

Fig. 58 (b) Epicarp of fruit is used to make utencils and muscical instrument

59

Lantana camara L.

Common name : English- Wild sage

Hindi- Raimuniya

Odia- Naguari

Family : Verbenaceae

Habit & Distribution

It is a shrub (Fig. 59 a), commonly found in roadsides, wasteland, damp site etc. It is distributed throughout India.

General Characters

The plant is an evergreen shrub grows up to 6 ft height and may spread 8 ft in width. The leaves are ovate-oblong, about 2-5 inches long with rounded tooth edges and a textured surface. Stem and leaves surfaces are covered with rough hairs and emit an unpleasant aroma when crushed. Flowers usually white-pink or orange-yellow in colour and are held in clusters that are typically 1-2 inches across. Fruits are small, globose, fleshy, purplish black berries after ripe. The plant bears flowers and fruits in all round the year.

Part Used as Play Material

The fruits of the plant is rounded, with plain surface and arranged in clusters that just look like playing bullets of guns. Children use the seeds as bullets in different handmade guns as fun. A hollow bamboo stick is used by the children, in to which they insert the fruits, in line followed by pressing the inserted fruits with a light stick. As a result, fruit comes out at the opposite end making sound like cracker (Fig.59 b). The seeds are also used for counting purpose.

The fruits are also eaten by the children as berries, when it ripens and it tastes sweet.

Fig. 59 (a) *Lantana camara* L.

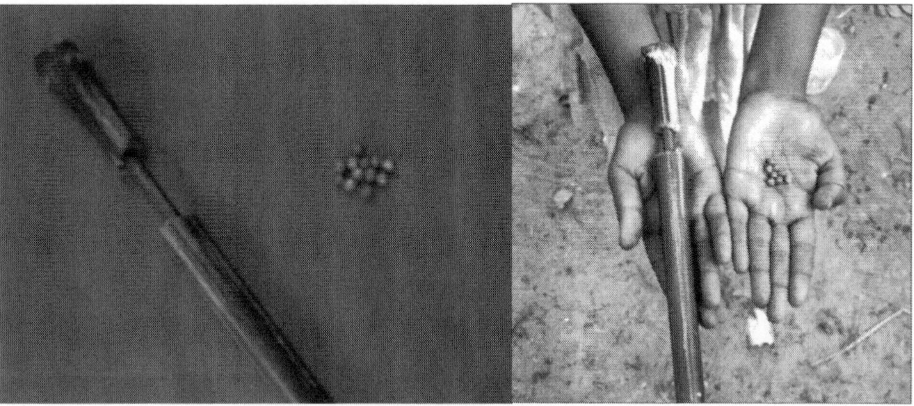

Fig. 59 (b) Children use the fruits as Bullets

60

Luffa acutangula (L.) Roxb.

Common name : English- Angled luffa

Hindi- Karvitori

Odia- Pita taradi

Family : Cucurbitaceae

Habit & Distribution

The plant is a climber (Fig. 60 a), widely cultivated and also found in wild. It is cultivated throughout India.

General Characters

It is an annual climber with smooth and angled stem. The leaves are 5-7 lobed or angled, 7.5-15 cm across and covered with short hairs. Flowers are yellow in colour with male and female in the same axil. The fruits are ribbed and cylindricaly shaped and 1-2 feet in length. The plant bears flowers and fruits during August-November.

Part Used as Play Material

The tip of the dried fruits are removed and cleaned and dried. It is used as a rotating body on the ground surface for play purpose (Fig.60 b).

Fig. 60 (a). *Luffa acutangula* (L.) Roxb.

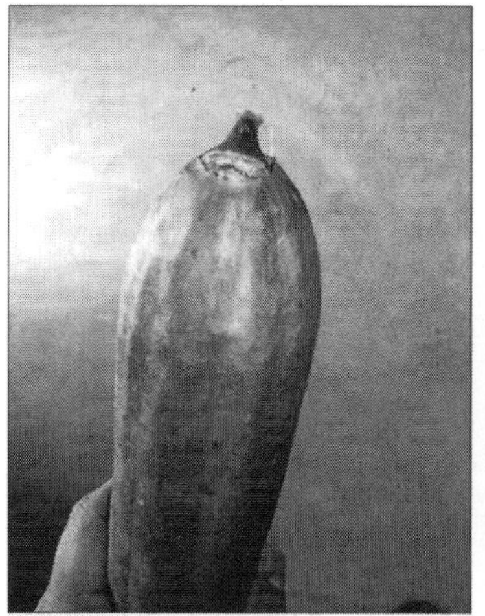

Fig. 60 (b) Fruits are used for makling Chakri (wheel)

61

Mangifera indica L.

Common name : English- Mango
Hindi- Aam
Odia- Amba

Family : Anacardiaceae

Habit & Distribution

It is a tree (Fig. 61 a) grows wild in valleys and widely cultivated throughout tropical India.

General Characters

The plant is a large, erect and ever green tree gowing up to 30-80 ft in height with a broad and rounded canopy. The leaves are variable in size and shape, usually lanceolate, entire, acute to acuminate with 10-20 cm in size. The flowers are greenish-yellow, small and polygamous with 5 lobed calyx which are ovate-oblong or elliptic in shape and remain spreading. There is a great variation in the size, colour and flavour of the fruits. They may be nearly round, oval, ovoid-oblong or somewhat kidney shaped and usually yellow in colour when ripe. The plant bears flowers during January-March and fruits during April-May.

Plant Part Used as Play Material

The leaves are entire, long with a midrib. Children use leaves to make some craft items like small vanity bags, by folding the leaves to make bag and the mid rib of the leaves as handle of the bag as their play material. They also use mango leaves for making letters (A, B, C, D) by cutting it into different sizes. Children also use the leaves as currency notes while playing different games. The leaves are also used in docio-religious functions.

Fruits: After eating the pulpy part of the fruit, the remaining seed of the fruit is dried. The dried fruit is rubbed at one side in an angular manner. Children use it as a whistle and blow to make sounds (Fig.61 The fruits have nutritional value and rich with sugar and vitamins.

Fig. 61 (a) *Mangifera indica* L.

Fig. 61 (b) Leaves used as currency (notes) and seed used to make whistle

62

Martynia annua L.

Common name : English- Tiger's claw/Devil's claw

Hindi- Ulat kanta

Odia- Bagha nakhi

Family : Martyniaceae

Habit & Distribution

The plant is a herb (Fig. 62 a), commonly occur in roadsides, waste lands, and also naturalized. It is widely distributed in India.

General Characters

It is an erect, often branched herb that grows up to 2 m in height. Leaves are kidney shaped to circular, mostly 6-15 cm wide, both surfaces are equally hairy, margins with shallow lobes and leaf stalk is 9-14 cm long. Flowers are bell shaped purplish white with dark purple markings. Fruit is capsule, more or less ovoid, 2 valved, with the style splitting into 2 curved, lignified and spiny processes with curved claw of about 1 cm. The plant bears flowers and fruits during August-January.

Part Used as Play Material

The fruit with a curved spiny claw, which looks like the claws of a beer. Children use these fruits for making fun and they play by pinching each other with the spiny fruits (Fig.62 b).

Fig 66 (a): *Martynia annua* L.

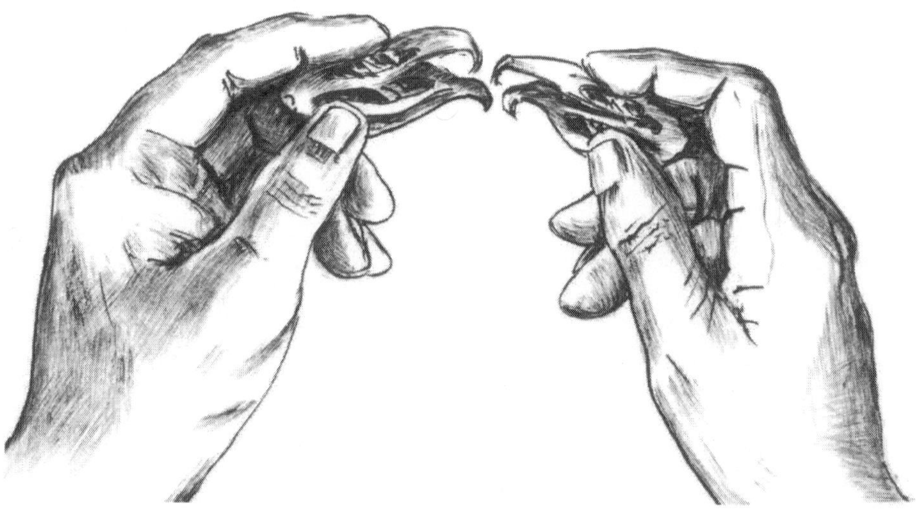

Fig 66 (b): Fruits are used as play material as it carries spines and looks like tigers claw

63

Mimosa pudica L.

Common name : English- Touch me not/sensitive plant

Hindi- Lajwanti

Odia- Lajakuli

Family : Mimosaceae

Habit & Distribution

The plant is a common weed mostly found in roadsides, waste places, damp sites, agricultural lands etc. It is widely distributed throughout the better parts of India. (Fig. 63 a)

General Characters

It is a small prostrate or ascending herb with prickly stems and grows up to a height of 50 cm with a spread of 30 cm. The leaves are bipinnate, fern like and pale green, closing when disturbed. The flowers are pale pink or purple in colour, round to ovoid heads (8-10 mm in diameter) and the flower heads arise from the leaf axils. The plant bears flowers and fruits during August-May.

Part Used as Play Material

It is a very wonderful and curious plant. It is very sensitive to touch. The leaves of the plant have folding and recoiling property. The leaves fold inward automatically when touched. After sometime it gets back to original position. For this property, children make fun with the leaves by touching it by leg or stick (Fig.63 b). The plant is known as Lajakuli (touch me not). It has also medicinal value.

Fig. 63 (a) *Mimosa pudica* L.

Fig. 63 (b) Children play with the leaves as it is sensitive and closes when they are touched

64

Mimusops elengi L.

Common name : English- Spanish cherry

Hindi- Maulsari

Odia- Baula

Family : Sapotaceae

Habit & Distribution

The plant is a small tree (Fig. 64 a), frequently planted elsewhere also in gardens. It is distributed throughout India.

General Characters

It is a small tree with small shiny thick narrow, pointed leaves, straight trunk and spreading branches. The leaves are elliptic or oblong, 6.2-10 cm long. Flowers are small star shaped, white, yellowish white in color with a crown rising from the centre. The fragrance of flowers is very graciously scented. The fruit turns yellow-orange when ripens and are eaten fresh. The plant flowers during April-May and bears fruits during August-September.

Part Used as Play Material

The flowers are white in color and very much scented. Children love to keep it with them in their bags and pockets. The flower has also ornamental use. Girls use the flowers to make different ornaments (imitated) like brasslate, ear rings to wear as it is very much scented and star shaped. (Fig. 64 b)

The fruits are also eaten by the children as berries, because of its good taste and color.

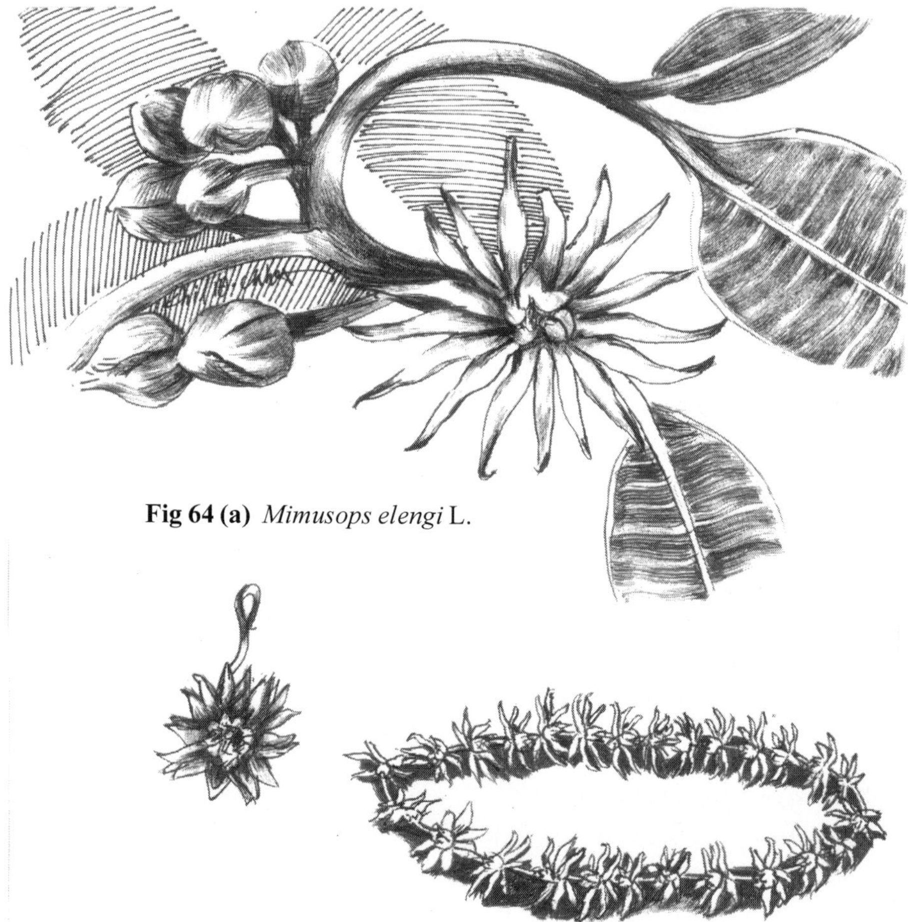

Fig 64 (a) *Mimusops elengi* L.

Fig 64 (b) Flowers are used for making Ornaments

65

Musa paradisiaca L.

Common name : English - Banana

Hindi - Kela

Odia - Kadali

Family : Musaceae

Habit & Distribution

The plant is a large herb (Fig. 65 a), indigenous to Odisha, Bihar and Eastern Himalayas. The plant is cultivated throughout India.

General Characters

Banana is a tropical small size tree, with large leaves with overlapping bases from the trunk. The leaves are oblong and 1.2-1.5 m in length. The stem reaches a height of about 10-30 feet. Flower develop from the centre of crown. Only female flowers develop into a banana fruit that vary in length from 4 to12 inches. Each banana plant bears fruits only once. The plant bears flowers and fruits in most parts of the year.

Part Used as Play Material

The stem sheath of banana plant is cylindrical, long and white in colour. It is also very much light to handle, soft and plain, by which it is easy to cut for making different shapes. Children use it to make boats of different models for playing. It is soft and light, that can easily float on water.

The leaf stalk is splited into two parts and used to make sound like craker. The dried midrib is used to make toys like cars, boates etc. The spathe is brown in colour and boat shapes. Children use this spathe as slipper for walking (Fig.65 b).

The plant carries lots of economic value. The fruit is eaten, both in raw and ripen form and is rich in potassium. The leaves of the plant are used as plates, and sometimes as firewood when dried.

Fig. 65 (a) *Musa paradisiaca* L.

Fig. 65 (b) Stem sheath used to make boat, Leaf stalk used to make drum stick and Spathe used as slipper in games

66

Neolamarckia cadamba (Roxb.) Bosser.

Common name : English- Bur flower

Hindi- Kadamb

Odia- Kadamba

Family : Rubiaceae

Habit & Distribution

It is a tree (Fig. 66 a), seen growing widely in deep moist, alluvial sites and often in forests along river banks. It is widely distributed throughout India.

General Characters

The plant is a large straight deciduous tree that reaches to 45 m height. The leaves are large simple, elliptic-oblong, entire and 13-32 cm in length. Flowers are orange colored, small, in dense, globose heads that appear like solid, hairy orange balls. The fruits are small capsules, packed closely together. It bears flowers on May-July and fruits during August-October.

Part Used as Play Material

The flowers are rounded, soft, scented and very attractive reddish yellow color. Therefore, children use the flowers as ball to play different games like volley (Fig 66 b).

Fig 64 (a) *Neolamarckia cadamba* (Roxb.) Bosser.

Fig 64 (a) Flowers are used as ball to play cricket/volly

67

Nicandra physalodes (L.) Gaertner

Common name : English- Shoo-fly plant

Hindi- Popti

Family : Solanaceae

Habit & Distribution

The plant is a herb (Fig. 67 a), commonly found in fallow land, naturalized in waste places and near dwellings. It is distributed throughout India.

General Characters

It s an erect annual plant reaches up to 3-8 feet in height and are about half as wide. The leaves are large, alternate, reaching up to 1 feet long and somewhat resembles *Datura* leaves. Flowers are pale blue or blue purple with white centre. Fruits are berry, brown or yellow when matures, up to 2 cm in diameter. The plant flowers and bears fruits during October-February.

Part Used as Play Material

The fruit of the plant is very light. Children use the fruits as cracker by filling it with air from mouth, followed by breaking it on their forehead (Fig.67 b).

Fig. 67 (a) *Nicandra physalodes* (L.) Gaertner

Fig. 67 (b) Children use the fruits as crackers

68

Nymphaea pubescens Willd.

Common name : English- Water lily

Hindi- Kanval

Odia- Kain

Family : Nymphaeaceae

Habit & Distribution

The plant (Fig. 68 a) is an aquatic perennial herb, commonly found in lakes and ponds. It is distributed throughout the warmer parts of India.

General Characters

It is an aquatic plant having erect perennial root stocks that anchor it to the mud in the bottom of the water bodies. Leaves vary from being elliptic-oval to almost circular 15-26 cm across, fine hairs on upper side and always remain at floating stage. Flowers are mostly white, occasionally pink or red reaching up to 15 cm in diameter. The plant flowers and bears fruits throughout the year.

Part Used as Play Material

The flower is large in size with attractive colour and having a long stalk. Children use the flowers for various decoration purposes. It is specially used by the girls for making ornaments by folding its long stalk.(Fig. 68 b)

Fig 68 (a) *Nymphaea pubescens* Willd.

Fig 64 (a) Flowers are used for decoration as well as ornamental purpose

69

Alangium salviifolium (L.f.) Wangerin

Common name : English- Sage Leaved Alangium

Hindi- Ankol

Odia- Ankula/Dhalanga

Family : Cornaceae

Habit & Distribution

The plant is a tree (Fig. 69 a), commonly found in dry regions in plains and low hills and also found on roadsides. The plant is widely distributed throughout India.

General Characters

The plant is a tall thorny tree native to India. It grows to a height of about 3 to 10 meters. The bark is ash colored, rough and faintly fissured. The leaves are elliptic oblong or elliptic lanceolate. The flowers are greenish white, fascicled, and axillary. The berries are ovoid, ellipsoid, glabrous, smooth and violet to purple in colour. The plant flowers and bears fruits during February -June.

Part Used as Play Material

The leaves are elliptic and leathery. Children play with the leaves as their currency notes for fun making and make bundles of these leaves as note bundles (Fig.69 b).

Fig. 69 (a) *Alangium salviifolium* (L.f.) Wangerin

Fig. 69 (b) Leaves are used as money (currency) for playing

70

Oryza sativa L.

Common name : English-Rice

Hindi-Chawal

Odia-Dhana

Family : Poaceae

Habit & Distribution

The plant (Fig. 70 a) is a annual herb, and several varieties of this plant are largely cultivated all over the world, especially in Asian countries.

General Characters

It is a semi aquatic monocot plant. The plant grows up to 3-5 ft in height. It has long, slender leaves, 50-100 cm long and 2-2.5 cm broad. Flowers are produced in branches arching to pendulous inflorescence 30-50 long. The edible seed is a grain, 0.5-1.2 cm long and 2-3 mm thick. The plant flowers and bear fruits during September-January and also in April-May.

Parts Uses as Play Material

After maturity and ageing, the rice stalks/stem becomes yellow in colour. Children use it as a straw to suck water by dipping one end into the water and the other part in their mouth for fun. Further, they make fun by making bubbles by dipping the stalk into foam (Fig.70 b). Besides these, children make dolls and puppets from the straw, as well.

The rice straw is an agricultural by product, after the grain and chaff are removed. It has many uses, including fuel and fodder, thatching and basket making.

Fig.70 (a) *Oryza sativa* L

Fig.70 (b) Children use dry stalk as straw to suck liquids and to make bubbles from foam

71

Pandanus odorifer (Forssk.) Kuntze

Common name : English- Kewda/Screw pine

Hindi- Gagan dhul

Odia- Kia

Family : Pandanaceae

Habit & Distribution

It is a shrub (Fig. 71 a), wild in deltaic swamps of Mahanadi and other sandy coastal areas. Seen both in wild and cultivated in India.

General Characters

It is a bushy shrub, reaches up to 3-4 m in height. The leaves are closely spiral, spiny and leathery with fragrance. Flowers are pedicellate with stamens spirally arranged on the floral axis. Flowers have a sweet and perfumed odor.

Part Used as Play Material

The flowers are scented and have a pleasant flavour similar to rose. Children (especially girls) usually use the flowers to decorate their hairs (Fig.71 b).

Fig.71 (a) *Pandanus odorifer* (Forssk.) Kuntze

Fig.71 (b) Flowers are used to keep in bags by children and for decoration of hair

72

Peltophorum pterocarpum (DC.) K. Heyne

Common name : English-Yellow flamebouyant

Hindi- Peela gulmohar

Odia- Radhachuda

Family : Fabaceae

Habit & Distribution

It is a tree (Fig. 72 a), frequently planted in gardens, road sides etc. and widely distributed throughout India.

General Characters

The plant is a deciduous tree growing up to 15-25 m height with a trunk diameter of about 1m. The leaves are bipinnate, 30-60 cm long, with 16-20 pinnae, each pinna with 20-40 oval leaflets. The flowers are yellow, 2.5-4 cm diameter, produced in large compound racemes up to 5-10 cm long and 2.5 cm broad, red at first, ripening black and contain 1-4 seeds. The plant flowers during March-May and bears fruits during April-December.

Part Used as Play Material

The petals of the flowers of this plant are soft and yellow in colour that looks quite attractive. Children use the petals as artificial nails by joining it with water gum or other liquid as their nails artificially. The nails look long and colorful (Fig.72 b).

Stamens: The stamens are used for game in which the strength of stamen is considered. The anthers are crossed with each other by the children and are pulled in opposite directions for making fun.

Fig. 71 (a) *Peltophorum pterocarpum* (DC.) K. Heyne

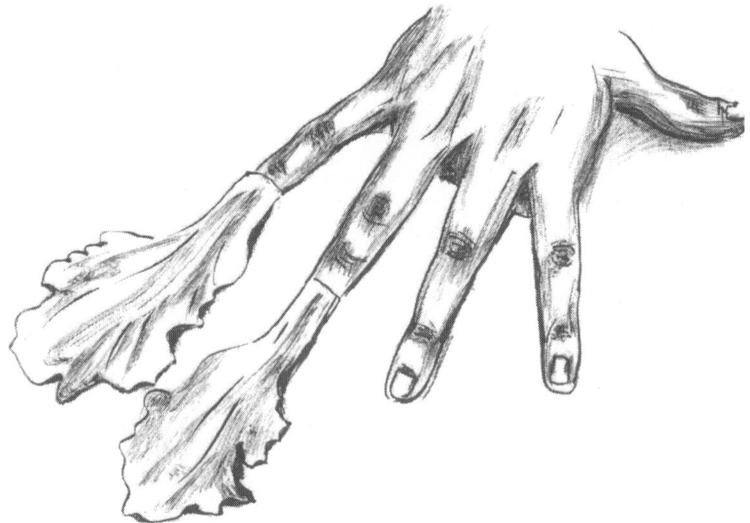

Fig.71 (b) Petals used to make artificial nails

73

Phoenix sylvestris (L.) Roxb.

Common name : English- Date palm

Hindi- Khajur

Odia- Khajuri

Family : Arecaceae

Habit & Distribution

The plant is a tree (Fig. 73 a). It is commonly, wild or cultivated throughout India. It thrives from the plains to the coast in low-laying waste lands, scrub forest and water logged areas.

General Characters

It is a moderate sized dioecious tree which grows up to 4-10 m. in height and 30 cm diameter with a large crown and rough trunk covered with persistent leaf bases. The leaves are 3 m. long, gently recurved on 1 m petioles with spines near the base. The leaf crown grows to 5 m wide and 7-10 m tall containing up to 100 leaves. The inflorescence grows to 1 m with white flowers. The single seeded fruit ripens to orange-yellow. The plant flowers and bears fruits during May-October.

Part Used as Play Material

Children use the leaves to make different craft items like handfans, mats by arranging the leaflets in horizontal and vertical manner in a series. It is also used to make several decorative, play materials and ornaments (Fig.73 b).

The fruits of the plant has also food value and eaten as berries.

Fig.73 (a) *Phoenix sylvestris* (L.) Roxb.

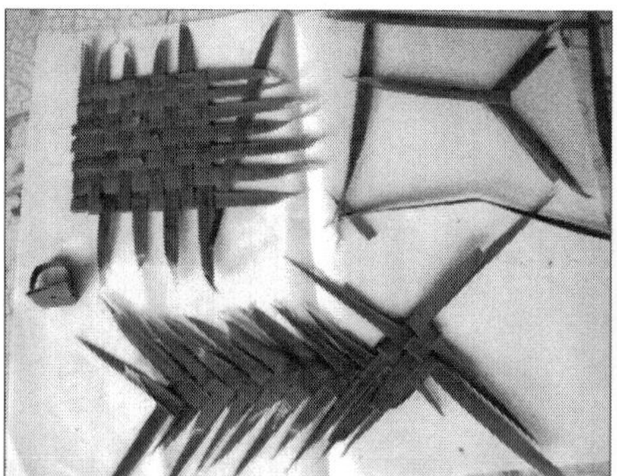

Fig.73 (b) Leaves are used to make different craft items, ornaments and toys

74

Plumbago zeylanica L.

Common name : English- Wild leadwort

Hindi- Chitrak

Odia- Ogni

Family : Plumbaginaceae

Habit & Distribution

The plant is a shrub (Fig. 74 a), frequent in open places, scrub forests and among bushes. It is widely distributed throughout India.

Gerneral Characters

The plant is an evergreen shrub that reaches about 6 feet in native. The leaves are dark green, spreading and ovate in shape. Flowers are white, terminal, often panicled and grow as dense racemes. Individual flowers are up to ½ inch across. The plant flowers and bears fruits during September-April.

Part Used as Play Material

The flowers are very light in weight and white in colour. Children use the flowers to decorate their eyelashes for fun because the flowers can easily get attach with the eyelash. (Fig. 74 b)

The plant has also medicinal importance, used as medicine for dysentery, piles and itching diseases by tribals.

Fig.74 (a) *Plumbago zeylanica* L.

Fig.74 (b) Children use the flowers to decorate eyelashes

75

Polyalthia longifolia (Sonn). Thwaites

Common name : English- False Asoka

Hindi- Ashok

Odia- Debadaru

Family : Annonaceae

Habit & Distribution

The plant (Fig. 75 a) is usually planted in gardens, roadside avenues and also cultivated throughout India.

General characters

The plant is a tall evergreen tree having slender and glabrous branches, growing up to a height of 25 feet. The leaves are narrowly lanceolate, glossy green, long with attractive wavy edges. Flowers are yellowish-green, star like structures that give the tree a peculiar hazy appearance. Seeds are smooth and shining. The plant flowers in March-April and bears fruits during July-August.

Part Used as Play Material

The leaves of these plants are narrow and having wavy margins, which look very attractive. Children usually use the leaves for decoration purposes in different festivals (Ganeshpuja, saraswati puja) by making necklace (mala)

While making small mud houses for play, children use the branches with dense leaves for thatching (Fig.75 b).

Fig.75 (a) *Polyalthia longifolia* (Sonn.) Thwaites

Fig.75 (b) Leaves used for decoration purpose and branches used to built small houses for playing

76

Pseudobombax ellipticum (Kunth) Dugand

Common name : Shaving brush tree

Family : Malvaceae

Habit & Distribution

The plant is a large tree (Fig. 76 a), usually seen in open forests often scattered through fields and over plains, besides on open rocky hillsides. It is distributed throughout Indian.

General characters

It is large deciduous tree, which grows up to 30-40 feet in height. The leaves at first are bright red turning fine green as they mature. Flowers are having silky rose-pink stamens topped with yellow pollen. The wood is interesting as well, showing strippings of greens, yellows, brown and white. The plant flowers during October-February.

Parts Used as Play Material

The flowers have numerous hairy and silky stamens with pink color, which appears like a shaving brush. Children use the flowers as shaving brush as play material to make fun (Fig.76 b).

Fig.76 (a) *Pseudobombax ellipticum* (Kunth) Dugand

Fig.76 (b) Childern use the flower as shaving brush

77

Ricinus communis L.

Odia- Jada/Gab

Common name : English- Castor bean

Hindi- Arandi

Odia- Jada/Gab

Family : Euphorbiaceae

Habit & Distribution

The plant is a shrub or small tree (Fig. 77 a), often found in wastelands, fields, gardens and near villages. Also cultivated as a cash crop. It is widely distributed throughout the dry parts of India.

General Characters

The plant is a fast growing, sucking perennial tree that reaches the size up to 12 m. The leaves are 15-45 cm long, stalked, alternate and palmate with 5-12 deep lobes. The flowers are borne in terminal panicle like inflorescences of green or shades of red monoecious flowers without petals. The fruit is spiny, greenish capsule containing large and oval bean like seeds. The plant flowers and bears fruits during most parts of the year.

Part Used as Play Material

The fruits are oval shaped and spiny. Children use it as weapon as their play material (Fig. 77 b). They play with it by throwing the fruits at each other. The plant produces castor oil. The young buds are poisonous to cattles.

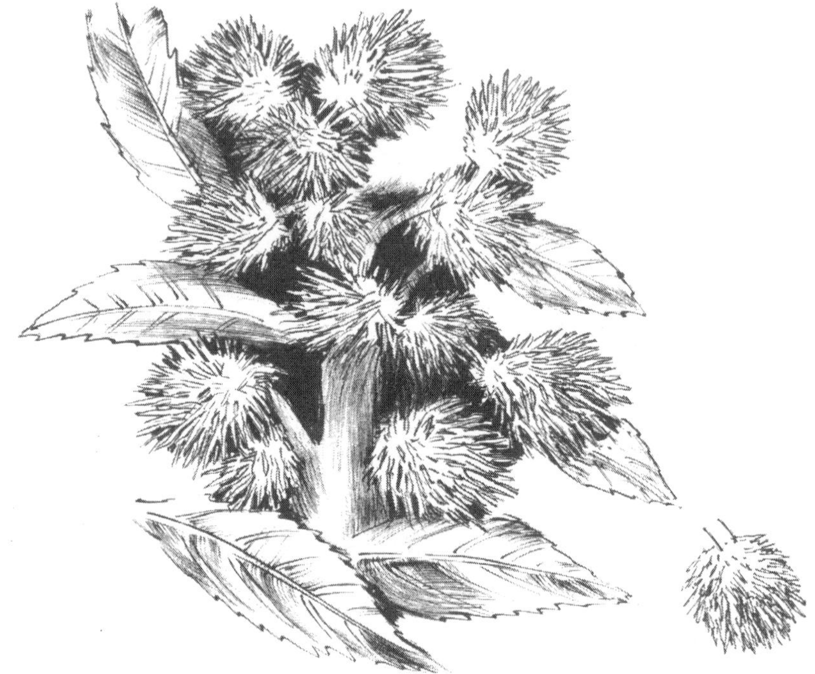

Fig.77 (a) *Ricinus communis* L.

Fig.77 (b) Fruits are used as weapon by children

78

Rosa indica L

Common name : English- Rose

Hindi-Gulab

Odia- Golap

Family : Rosaceae

Habit & Distribution

It is a shrub (Fig. 78 a), commonly planted in gardens and also cultivated.

General Characters

The plant is a woody, perennial shrub. The leaves are borne alternately on the stem. In most species, they are 5-15 cms long, elliptic-lanceolate, pinnate with 3-11 leaflets. Flowers are usually white or pink, though in a species yellow or red with 5 cm across. The aggregate fruit of the rose is a berry like structure called a rosehip. The plant flowers during February-March.

Part Used as Play Material

The flowers have got attractive colors available in different shades like pink, red or white with a special fragrance. Children (especially by the girls) usually use the petals of this flower for coloring nails. It has ornamental use as the girls use rose for decorating their hairs for its beautiful color and attractive scent (Fig. 78 b).

Fig.78 (a) *Rosa indica* L

Fig.78 (b) Flowers are used for nail art or decoration by children

79

Ruellia tuberosa L.

Common name : English- Fever root
Hindi- Ruwel

Family : Acanthaceae

Habit & Distribution

It is a herb (fig. 79 a), often found in wastelands, cultivated land, waysides etc. sometimes in gardens also. The plant is widely naturalized in India.

General Characters

The plant is a tropical perennial herb, growing up to a height of 6.5 inches with hairy stem. The leaves are simple, opposite and elliptic-obovate with 4.10-1.8-4 cm in size. Flowers are blue or violet with 3-4 cm in long. The ripe fruits in a pod carry 7-8 seeds each in it. The plant flowers and bears fruits during June-February.

Part Used as Play Material

The fruit of the plant is a pod, which bursts open with a sound when they come in contact with water and the black seeds are hurdled away. Children like to play with the dry pods that pop when rubbed with water (Fig.79 b). For this, its name as popping pod, duppy gun and cracker plant comes from the fact.

Fig.79 (a) *Ruellia tuberosa* L.

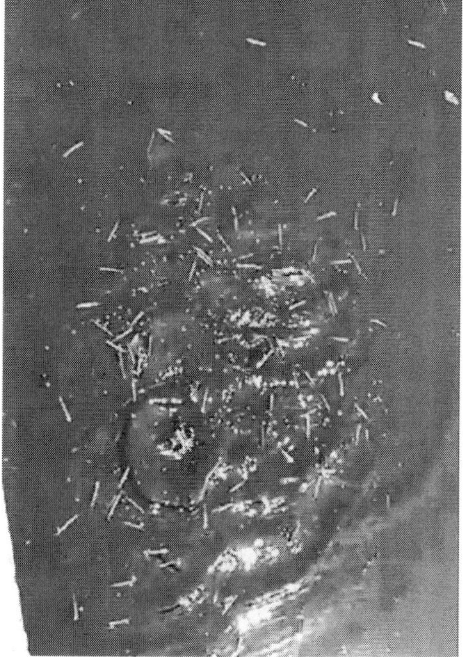

Fig.79 (b) Children put the seeds in water to note the cracking

80

Saccharum officinarum L.

Common name : English- Sugarcane

Hindi- Ganna

Odia- Akho

Family : Poaceae

Habit & Distribution

It is a tall herb (Fig. 80 a), widely cultivated in Odisha and also throughout the hotter parts of India.

General Characters

The plant is a giant grass. The stem is erect, jointed, 3-5 m tall, 2-3 cm thick with plenty of juice. Leaves are elongate, rigid, mostly 4-6 cm wide, with a very thick midrib. The flowers are white, appear in plume like panicles, 20-60 cm long. The plant flowers and bears fruits during November-January.

Part Used as Play Material

The stalk of the inflorescence is cut into pieces about 5 inches in length and made into bundles of ten. Children keep ten bundles in a small basket and use those for counting and simple counting arithmetic (Fig.80 b).

The plant has also great economic value as it is a source of sugar production.

Fig. 80 (a) *Saccharum officinarum* L.

Fig. 80 (a) Inflorescence stalk are used for general counting /multiplication/addition etc

81

Solanum viarum Dunal

Common name : English- Tropical soda apple

Odia- Denga bheji/Bheji baigana

Family : Solanaceae

Habit & Distribution

It is a perennial shrub (Fig. 81 a), is commonly found in grassland, thickets and disturbed places such as roadsides and river banks.

General Characters

The plant is an erect perennial shrub, reaches up to 50-150 cm high with shortly pubescent stems and branches with recurved prickles up to 5 mm long. The leaves are broadly ovate up to 20 cm long and 15 cm wide, generally dark green in colour, and glossy above. The flowers are white, 1.5 cm across arranged in clusters. The fruit is a globose berry, mottled green when young and yellow at maturity, with 2-3 cm across. The plant flowers and bears fruits throughout the year.

Part Used as Play Material

The fruit is rounded in shape. Children use it to play 'goti/guli/bati' game (Fig.81 b).

Fig.81 (a) *Solanum viarum* Dunal

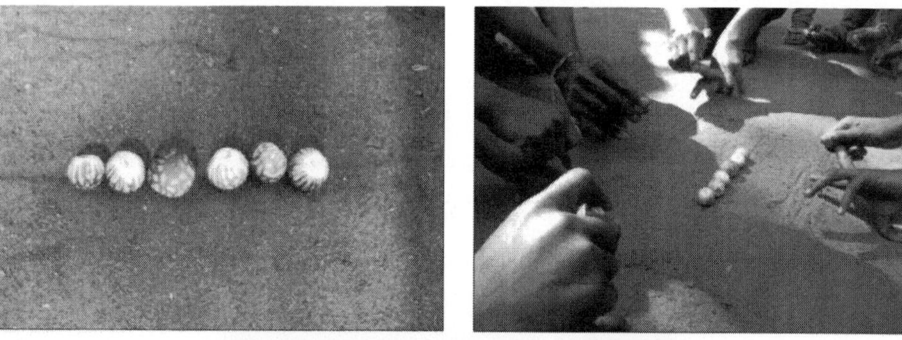

Fig.81 (b) Fruits are used as 'gottiyan' to play games like 'guli-bati'

82

Tagetes erecta L.

Common name : English- Marigold
Hindi- Genda
Odia- Gendu

Family : Asteraceae

Habit & Distribution

It is a herb (Fig. 82 a) commonly cultivated in gardens. It is a very popular flowering plant commonly found in India and other countries.

General Characters

The plant is a very popular aromatic annual herb with 50-100 cm in height. The leaves are pinnately divided, in to oblong or lanceolate. Flowers are yellow, orange in colour, and large head with 3-5 inches across. The plant flowers mainly during cold season.

Part Used as Play Material

The flowers are round, yellow or orange in colour. Further, the flowers are very light in weight. Children use it as a ball to play different games such as volley, badminton, tennis etc (Fig.82 b).

Fig.82 (a) *Tagetes erecta* L.

Fig.82 (b) Flower used as ball and microphone.

83

Tamarindus indica L.

Common name : English- Tamarind

Hindi- Imli

Odia- Tentuli

Family : Fabaceae

Habit & Distribution

The plant is a large tree (Fig. 83 a), fairly common, planted or self sown. It is widely distributed throughout India, also cultivated and self-sown.

General Characters

It is a large tree which grows up to 30-50 m height with spreading crown up to 12 m in diameter. The leaves are up to 5 cm long and composed of numerous small leaflets. Leaflets are 10-20 pairs, small and close. The flowers are about 1 cm across, pale yellow with purple or red veins. Fruits are brown, short haired, sausage like, contain an acidic pulp that surrounds the seeds, extremely sour. The plant flowers during April-June and bears fruits during December-March.

Part Use as Play Material

The tamarind seed have a juicy pulp, covered over it. After eating the covered pulp, the seeds are collected, which are hard and deep red to brown in colour. Children use these seeds to make 'gottiyan' for playing different games like 'ludo' or drawing squares on the floor (Fig.83 b). These seeds are also used for counting numbers besides addition, subtraction etc. by the students at primary level. Children also eat the unripe tamarind fruit, collected by them by climbing the tree i.e. as part of their fun.

The leaves of the plant also taste sour. Children treat the leaves as treat vegetable while playing with kitchen set. Besides the nutritive value of the fruit, the plant is used as firewood.

Fig.83 (a) *Tamarindus indica* L.

Fig.83 (b) Seeds are used to make 'gottiyan' for playing ludo, and leaves are used as food while playing in kitchen-set.

84

Tectona grandis L.f.

Common name : English- Teak
Hindi- Sagun
Odia- Sagwan

Family : Verbenaceae

Habit & Distribution

The plant is a tree (Fig. 84 a), naturally confined to south-western part of Orissa, also widely planted all over the state. The plant commonly distributed in central India to Orissa, also widely planted.

General Characters

It is a large deciduous and very popular timber tree. Leaves of the tree are large, opposite with 30-60 cm long and 15-30 cm broad. The flowers appear in large numbers in lax clusters at the end of branches with white colour. The fruit is about 15 mm across, spongy and enclosed in persistent calyx. The plant flowers during July-August and bears fruit during December-April.

Part Used as Play Material

The leaves of the plant are large and give a brown to red color when rubbed. Children rub the leaves on their palm and enjoy with colorful palms (Fig.84 b).

Fig.84 (a) *Tectona grandis* L.f.

Fig.84 (b) Leaf is used for colouration of palms by rubbing

85

Tinospora cordifolia (Willd.) Miers.

Common name : English- Indian Tinospora

Hindi- Gulbel

Odia- Gulochi

Family : Menispermaceae

Habit & Distribution

It is a shrub/climber (Fig. 85 a), frequently found in forests and widely distributed throught tropical India.

General Characters

The plant is a climber, with stems about 5 cm diameters with light grey and papery bark. The leaves are about 7.5-14 cm long, broadly ovate or orbicular, deeply heart shaped at the base. Flowers are racemes which are 7-14 cm in long, greenish yellow in colour and arranged in 3+3 sepals in 2 layers, the outer one are small, the inner one are large. Fruits are red, globose to subglobose with 6-8 mm across. The plant flowers during August-December and bears fruits during March-May.

Part Used as Play Material

The plant is a climber; it has long aerial shoots spreading over the area of support. Children use these long shoots as telephone cords and make fun of it (Fig.85 b).

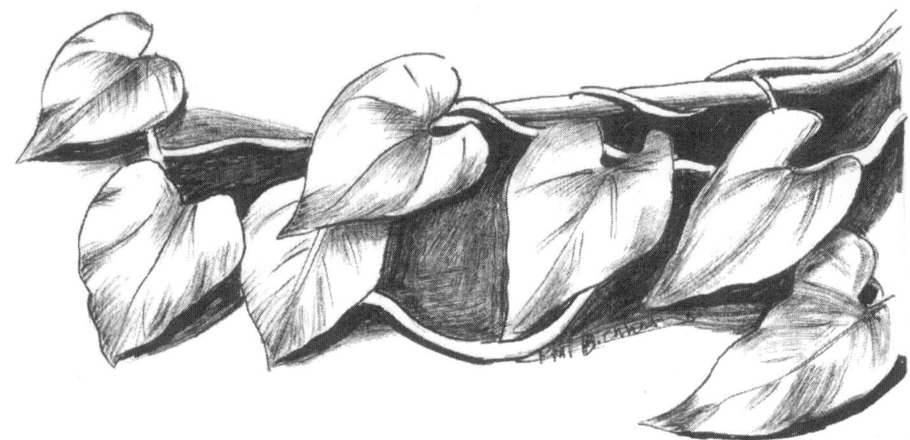

Fig. 85 (a) *Tinospora cordifolia* (Willd.) Miers.

Fig. 85 (b) Children use the stem as telephone cord

86

Trewia nudiflora L.r

Common name : English- False white teak

Hindi- Pindar

Odia- Jandakhai

Family : Euphorbiaceae

Habit & Distribution

The plant is a tree (Fig. 86 a), commonly found along riverbanks, evergreen and semi-evergreen forests. It is distributed throughout India.

General Characters

It is a large deciduous tree, growing up to 10-20 m tall, with a smooth grey bark. The leaves are oppositely arranged ovate, 7-17 cm long and are heart- shaped at the base and have tapering tips. Leaf-stalks are 3-7 cm long. Flowers are small and greenish. Male flowers are borne in 7-19 cm long racemes and are carried on slender stalks. Female flowers are borne singly on long stalks. Fruit is round, with 2-4 compartments, 2-4 cm in diameter, smooth or hairy, exocarp thick and fleshy. The plant flowers and bears fruits during December-March.

Part Use as Play Material

The fruit of the plant is round with a thick and fleshy exocarp. Children use the fruits for making wheels as their play material (Fig.86 b).

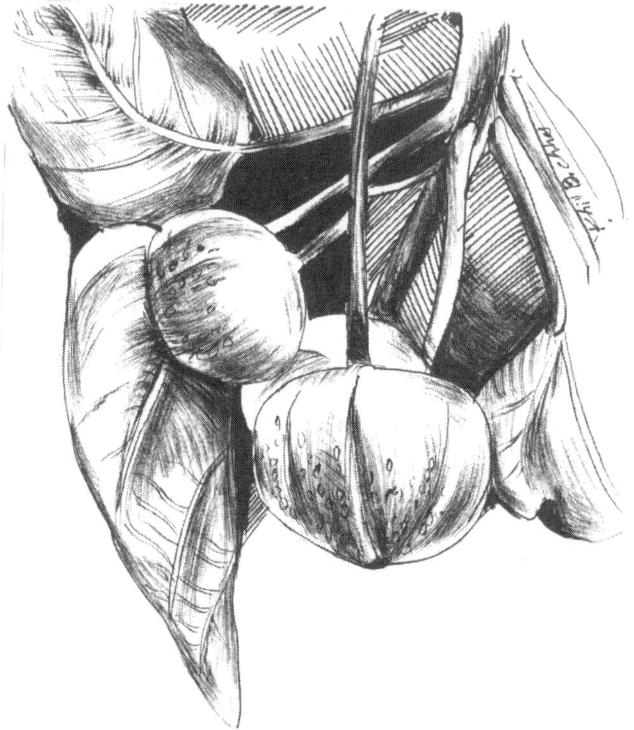

Fig. 86 (a) *Trewia nudiflora* L.

Fig. 86 (b) Fruits are used for making wheels as play material

87

Tridax procumbens (L.)

Common name : English- Coat buttons

Hindi- Khal-Muriya

Odia- Bisalya karani

Family : Asteraceae

Habit & Distribution

The plant is a herb (Fig. 87 a), usually found in dry locations, especially sandy and rocky sites like roadsides, railways, dunes and waste places. It is widely distributed throughout India.

General Characters

The plant is a perennial herb that stands about 30-60 cm in height and has slightly hairy stem. The leaves are ovate or lanceolate with toothed edges. Flowers are small, creamy or white in colour having 5 petals that are notched in the outer edges. The centre of the flower is yellow. The plant flowers and bears fruits all the year round.

Part Used as Play Material

The flowers of the plant are having white rays of petals with disc at centre that looks likes small rounded buttons. Children use the flowers to make 'tiara' to wear it on their head that looks beautiful (especially girls) and make fun out of it (Fig.87 b).

The plant has medicinal value, the leaf juice used as instant medicine for wounds.

Fig.87 (a) *Tridax procumbens* (L.)

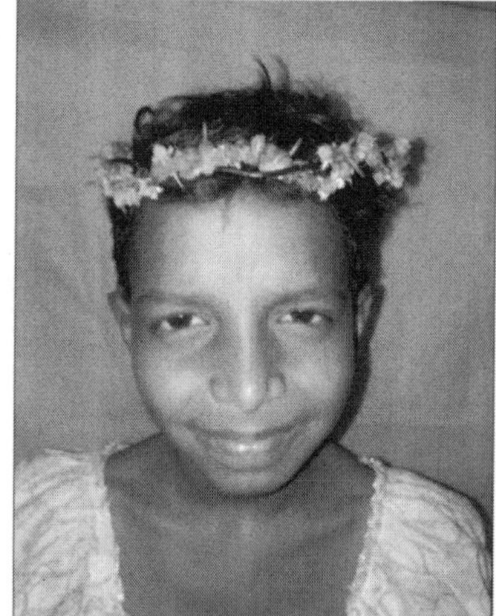

Fig.87 (b) Girls use the flowers to make tiara

88

Triumfetta rhomboidea Jacq.

Common name : English- Burr Bush

Hindi- Chikti

Odia- Badya chana pata/tatatatia

Family : Tiliaceae

Habit & Distribution

The plant is a herb (Fig. 88 a), fairly common weed, distributed throughout tropical and subtropical India.

General Characters

It is an crect woody herb, grows up to 30-90 cm. leaves are simple, alternate, broadly ovate to sub orbicular. Flowers are yellow, small and clustered on leaf axils. Fruit is a subglobose and about 3-4 mm in diameter, covered with 75-100 hooked spines of 1.0 to 1.5 mm long. The plant flowers and bears fruits during July-January.

Part Used as Play Material

Children use the fruits to make fun by throwing it to the hairs of each other, because it can easily stick to the hairs with its spines on its surface. Girls also use the fruits to decorate their hairs (Fig.88 b).

Fig.88 (a) *Triumfetta rhomboidea* Jacq.

Fig.88 (b) Children use the fruits for fun by throwing it to the hairs of each other.

89

Combretum indicum (L.) defilipps

Common name : English- Rangoon creeper

Hindi- Madhumalti

Odia- Madhumalati

Family : Combretaceae

Habit & Distribution

It is a climbing shrub (Fig. 89 a), cultivated widely in gardens and distributed throughout India.

General Characters

The plant is a twining climber blooming profusely throughout summer and can reach up to 70 feet in tropical climate. The leaves are elliptic-oblong and 7-15 cm long. Flowers are scented, pink-white to reddish-white in colour and grow in pendent racemes with a very fast rate. The plant flowers and bears fruits in most parts of the year.

Part Used as Play Material

Flowers of the plant are very scented, with attractive colours like pink-white to red-white, making a spectacular show. Children use it for making ornaments as well as different decoration purposes as their play material. They collect the flowers and insert the pedicel in to the flower whorl, which appear just like beautiful ring. They wear it in their fingers. Children also make necklace by joining many flowers by pedicels (Fig.89 b).

Fig.89 (a) *Combretum indicum* (L.) defilipps

Fig.89 (b) Flowers are used for making Ornaments (necklace, ring)

90

Xanthium strumarium L.

Common name : English- Common Cocklebur

Hindi- Chhota dhatura/Ghagra

Odia- Mendhaguli

Family : Asteraceae

Habit & Distribution

The plant is a herb (Fig. 90 a), found in cultivated lands, coastal dunes, water resources, roadsides, waste places and field edges. It is distributed throughout India.

General Characters

The plant has large and broad leaves, light and bright green in colour in an alternate pattern with irregular lobes. Stem turns maroon to black when mature, with hairy appearance. The flowers are white or green, numerous and covered with hooked bristles. The fruit is obovoid, 1.5-2.5 cm long and covered with hooked spines of 2-4 mm long. The plant flowers and bears fruits during August-September.

Part Used as Play Material

The fruits are covered with numerous hooks. Children make fun by throwing it at each other. The seed sticks to their hairs and dresses (Fig.90 b).

Fig. 90 (a) *Xanthium strumarium* L.

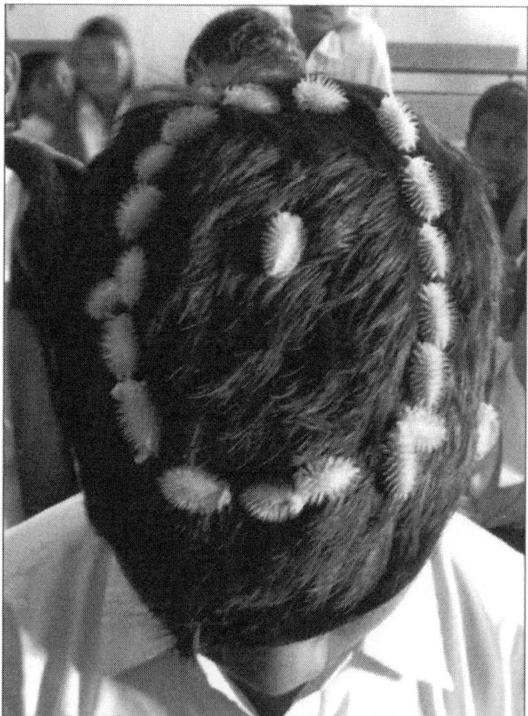

Fig. 90 (b) Children throw the seeds at each other and enjoy its sticking

List of Plants and their Part(s) Used for Fun and Games

S.N	Name of the Plant	Family	Common Name	Hindi Name	Parts Used As Play Material	Mode of use
1	*Abelmoschus esculentus* L. Moench	Malvaceae	Lady's finger	Bhindi	Fruit Stalk	Used to make horns
2	*Abrus precatorius* L.	Fabaceae	Indian liquorice	Ratti/Gunchi	Seed	Used for making necklace
3	*Abutilon indicum* (L.) Sweet	Malvaceae	Country Mallow	Kanghi	Fruit	Used for games & joy by joining fruits
4	*Acalypha hispida* Burm.f.	Euphorbiaceae	Red hot cat's tail		Flower	Used to make tails while playing different games.
5	*Achyranthes aspera* L.	Amaranthaceae	Prickly chaff flower	Chirchita	Leaf	Leaf is used as crackers
6	*Adenanthera pavonina* L.	Mimosaceae	Red Lucky Seed	Badi gumchi	Seed	Used to make necklace/ bracelets as it is deep red in colour
7	*Aeschynomene aspera* L.	Fabaceae	Sola, pith plant	Sola	Stem	Used to make boats and other craft items for playing as the it is the lightest wood among all the plants
8	*Aegle marmelos* (L.) Correa	Rutaceae	Wood apple	Bel	Fruit	Used as short foot for throwing
9	*Albizia saman* (Jacq.) Merr.	Mimosaceae	Rain Tree	Gulabi Siris	Chakunda	Used as money (coins) while playing
10	*Allium cepa* L.	Amaryllidaceae	Onion	Pyaz	Leaf	Used as straw to suck juice/water
11	*Areca catechu* L.	Arecaceae	Areca palm	Supari	Fruits, leaves	Fruit is used as ball and leaf branch used as lorry carrier
12	*Aristida setacea* Retz.	Poaceae			flower	Flowers stalk is used for making toy like snake.

S.N	Name of the Plant	Family	Common Name	Hindi Name	Parts Used As Play Material	Mode of use
13	*Artocarpus heterophyllus* Lam.	Moraceae	Jack fruit	Katahal	Leaves & fruits	Leaves are used for making utencils, whistles, tiaras and fruits are used for making toys like cow, horse etc.
14	*Azadirachta indica* A. Juss.	Meliaceae	Neem	Neem	Fruit, stem	Fruit used as marbles and stem used to make bows and arrows.
15	*Bambusa bambos* (L.) Voss	Poaceae	Bamboo	Bans	Stem	Used for making glass, pen stand, gun, flute, whistles, pushing car and young stem used as hockey stick
16	*Barleria prionitis* L.	Acanthaceae	Percupine Flower (Daskarenta/	Jhinti (H)	Fruits	Used as crackers by \ putting it into water
17	*Basella alba* L.	Basellaceae	Vine spinach	Poi	Fruit	Used for coloration of hands
18	*Bauhinia purpurea* L.	Caesalpiniaceae	Butterfly tree	Khairwal	Leaf	Used to make Containe of different shapes
19	*Bixa orellana* L.	Bixaceae	Annatto	Latkan/ Kumkum	Seeds	Seeds are used as colouration of palm and foots
20	*Bombax ceiba* L.	Bombacaceae	Red silk cotton	Shalmali	Stem	Used for making wheels and toys
21	*Borassus flabellifer* L.	Arecaceae	Palm	Taad	Fruit, leaf, stem	fruit used for making wheel, leaves for making watch, stem for making bat

S.N	Name of the Plant	Family	Common Name	Hindi Name	Parts Used As Play Material	Mode of use
22	*Bryophyllum pinnatum* (Lam.) Oken	Crassulaceae	Miracle leaf	Zakhm-haiyal	Leaf	Used for cleaning slates
23	*Butea monosperma* (Lam.) Taub.	Fabaceae	Bastard teak	Palash	Leaves, flower	Leaves used for making utensils (plates, bow) and flowers used for making coloured water (especially in holy)
24	*Caesalpinia bonduc* (L.) Roxb.	Caesalpiniaceae	Grey nicker	Kat-karanj	Seed	Used for making necklace and bracelets by girls and boys use it as marbles.
25	*Caesalpinia pulcherrima* (L.) Sw.	Caesalpiniaceae	Peacock flower	Guletura	Flower (petals)	Used for Making artificial nail
26	*Calotropis gigantea* R. Br.	Asclepiadaceae	Gaint milk weed	Akoa	Fruit, Flower bud	Fruit is used for making parrot and flower is broken to produce sound play
27	*Calotropis procera* (Aiton) Dryand	Asclepiadaceae	Rubber bush	Akada	Fruit, stem	Fruit used as ball and stem used as hockey stick.
28	*Carica papaya* L.	Caricaceae	Papaya	Papaya/Pateeta	Petiole/leaf stalk	Used for making pipe to pass water and to make whistle
29	*Cassia fistula* L.	Caesalpiniaceae	Golden shower	Sundaraj	Fruit	Used as sward to play
30	*Citrus aurantium* L.	Rutaceae	Orange	Santra	Spine,pulp	Spines used as injection while playing doctor game and pulp used as small bombs by exposing to fire

S.N	Name of the Plant	Family	Common Name	Hindi Name	Parts Used As Play Material	Mode of use
31	*Clerodendrum infortunatum* L.	Verbenaceae	Hill glory bower	Bhant	Flower(Stamen)	Stamens are used for playing by doing fight between 2 stamen to break their anther
32	*Cocos nucifera* L.	Arecaceae	Coconut	Nariyal	Shells, leaves and fruit	Shells used as utensils in kitchen set, telephone and making play material like tortoise, leaves are used to make watch, necklace, trumpet, hat, mat, baskets. Young coconut used for making cart
33	*Codiaeum variegatum* (L.) Rumph. Ex A. juss	Euphorbiaceae	Croton		Leaf	Used for making flower bookies
34	*Coix lacryma-jobi* L.	Poaceae	Adlay	Gurlu	Seed	Used as egg while playing kitchen game
35	*Colocasia esculenta* (L.) Schott	Araceae	Elephant ear taro	Kachu/Kachlu	Leaf	Used to capture dragonfly and play water bubble game
36	*Corchorus capsularis* L.	Malvaceae	White jute	Titapat	Stem	Use for making cigarette, straw/pipe as play material
37	*Cordia dichotoma* G.Forst.	Boraginaceae	Indian cherry	Lasora	fruit	Used as gum to prepare kites
38	*Crotalaria pallida* Aiton.	Fabaceae	Rattle pods	San	Fruit	Used as crackers
39	*Cucumis sativus* L. var. *hardwickii* (Royle) Gabaev	Cucurbitaceae	Wild cucumber	Khira	Fruit	Used for making toys like Chickes, horse etc.

S.N	Name of the Plant	Family	Common Name	Hindi Name	Parts Used As Play Material	Mode of use
40	*Cuscuta reflexa* Roxb.	Cuscutaceae	Giant dodder (Nirmuli)	Amar bel	Vines	Used as noodles while playing kitchen game
41	*Cynodon dactylon* (L.) Pers.	Poaceae	Bermuda grass	Dub	Shoot	Used for sweeping while playing
42	*Datura metel L.*	Solanaceae	Datura	Dhatura	Fruits	Used for punching one another by childrens to make fun as it carries spines.
43	*Dalbergia sissoo* DC.	Fabaceae	Rosewood	Shisham	Wood	Used for making bats
44	*Delonix regia* (Hook.) Raf.	Fabaceae	Gold mohar	Gul mohar	Seed pods, Stamen	Seed pods are used to prepare fan by children and Stamens are used for playing by doing fight between two stamen to break their anther
45	*Dillenia indica* L.	Dilleniaceae	Elephant apple/ouu	Chalta	Fruit	Used as short foot for throw
46	*Drypetes roxburghii* (Wall.) Hurus.	Euphorbiaceae	Lucky Bean Tree	Putija	Fruit	Used to make 'Natu' for playing
47	*Erythrina variegata* L.	Fabaceae	Indian coral tree	Pangara	Flower	Used for colouration of palm & decoration as it is deep red in colour
48	*Eucalyptus globulus* Labill.	Myrtaceae	Blue gum	Nilgiri	Bud	Operculum of stamen is used as tooth(canines) and dried ovary used as top

S.N	Name of the Plant	Family	Common Name	Hindi Name	Parts Used As Play Material	Mode of use
49	*Ficus benghalensis* L.	Moraceae	Banyan	Barh	Leaves, prop roots	Leaves are used to make utensils and umbrellas, prop roots are used as the swings.
50	*Ficus racemosa* L.	Moraceae	Clister fig tree	Goolar	Fruit	Used for making Cars
51	*Ficus religiosa* L.	Moraceae	Peepal tree	Peepal	Leaves	Used to make whistles, decoration, greeting cards and swings on branches
52	*Impatiens balsamina* L.	Balsaminaceae	Garden Balsam	Gul-mendi	Flower	Used for Colouration of Palms and decoration purposes.
53	*Ipomoea hederifolia* L.	Convolvulaceae	Scarlet morning	Lalpungli	Flower	Used for decoration of foreheads, also used as ornaments.
54	*Ipomoea parasitica* (Kunth) G. Doz	Convolvulaceae	Yellow-throated Morning Glory	Neelkalme	Leaves	Used for making different ornaments for playing
55	*Jasminum arborescens* Roxb.	Oleaceae	Royal jasmin	Chameli	Flower	Used for decoration of hair
56	*Jatropha gossypifolia* L.	Euphorbiaceae	Bellyache bush	Ratanjoti	Petiole, seeds	By breaking the petiole foam bubbles are prepared for playing, seeds are used to play king-queen game
57	*Jatropha integerrima* Jacq.	Euphorbiaceae	Spicy jatropa		Stem, flower	Stem used as making baskets and flower for ornamental pupose.

S.N	Name of the Plant	Family	Common Name	Hindi Name	Parts Used As Play Material	Mode of use
58	*Lagenaria siceraria* (Molina) Standl.	Cucurbitaceae	Calabash (Lau tumba)	Lauki	Fruit	Epicarp of fruit is used to make utencils and musical instrument
59	*Lantana camara* L.	Verbenaceae	Wild sage	Raimuniya	Fruits, flower	Bullets
60	*Luffa acutangula* (L.) Roxb.	Cucurbitaceae	Angled luffa	Karvitori	Fruits	Used for makling Chakri
61	*Mangifera indica* L.	Anacardiaceae	Mango	Aam	Leaves, seed	Leaves used as currency (notes) and seed used to make whistle
62	*Martynia annua* L.	Martyniaceae	Tiger's claw	Ulat kanta	Fruits	Used as play material as it contains spines & looks like1tiger's claw
63	*Mimosa pudica* L.	Mimosaceae	Shame plant	Lajwanti	Leaves	Children play with the leaf as it is sensitive & are closes when they touched
64	*Mimusops elengi* L.	Sapotaceae	Bullet-wood tree/ Baula	Maulsari	flower	Used for making Ornaments (brasslate, ear rings)
65	*Musa paradisiaca* L.	Musaceae	Banana	Kela	Stemsheath, Leaf stalk, spathe of inflorescence	Stem sheath used to make boat, Leaf stalk used to make drum stick and Spathe used as slipper in games
66	*Neolamarckia cadamba* (Roxb.) Bosser.	Rubiaceae	Bur flower	Kadamb	Flower	Used as ball to play cricket/volley
67	*Nicandra physalodes* (L.) Gaertner	Solanaceae	Shoo-fly plant	Popti	Fruit	Used as crackers
68	*Nymphaea pubescens* Willd.	Nymphaeaceae	Water lily	Kanval	Flower	Used for decoration as necklace

S.N	Name of the Plant	Family	Common Name	Hindi Name	Parts Used As Play Material	Mode of use
79	*Alangium salviifolium* (L.f.) Wangerin	Cornaceae	SageLeaved Alangium	Ankol	Leaf	Used as money (notes) for playing
70	*Oryza sativa* L.	Poaceae	Rice	Chaval	Dry stalks	Dry stalk used for fuel fooder and for thatching children mainly used the straw to drink and to make bubbles from foam.
71	*Pandanus odorifer* (Forssk.) Kuntze	Pandanaceae	Screw pine	Kewra	Flower	Used for decoration of hair as it is very much scented.
72	*Peltophorum pterocarpum* (DC.) K. Heyne	Fabaceae	Yellow flamebouyant	Peela gulmohar	Petals, stamens	Petals used to make artificial nails and anthers are crossed with each other and pulled opposite direction by children while playing
73	*Phoenix sylvestris* (L.) Roxb.	Arecaceae	Date palm	Khajur	Leaves	Used to make different craft items, ornaments and toys.
74	*Plumbago zeylanica* L.	Plumbaginaceae	wild leadwort	Chitrak	Flower	Used to decorate eyelashes
75	*Polyalthia longifolia* (Sonn). Thwaites	Annonaceae	False Asoka (Deb daru)	Ashok	Leaf, branches	Leaves used for decoration and branches used to built small houses for playing
76	*Pseudobombax ellipticum* (Kunth) Dugand	Malvaceae	Shaving brush tree		Flower	Used as shaving brush by children (Fruits)

S.N	Name of the Plant	Family	Common Name	Hindi Name	Parts Used As Play Material	Mode of use
77	*Ricinus communis* L.	Euphorbiaceae	Castorbean	Arandi	Fruit	Used as weapon by children
78	*Rosa indica* L.	Rosaceae	Rose	Gulab	Flower (petals)	Used for nail art or decoration by children
79	*Ruellia tuberosa* L.	Acanthaceae	Fever root	Ruwel	Pod	Fruits are used as craker by putting in water
80	*Saccharum officinarum* L.	Poaceae	Sugarcane	Ganna	Inflorescence stalk	Used for general counting/ multiplication/addition etc.
81	*Solanum viarum* L.	Solanaceae	Tropical soda apple		Fruit	Used as 'gottiyan' to play games like 'guli-bati'.
82	*Tagetes erecta* L.	Asteraceae	Marigold	Genda	Flower,seed	Flower used as ball
83	*Tamarindus indica* L.	Fabaceae	Tamarind	Imli	Seeds, leaves	Seeds are used to make 'gottiyan' for playing ludo, and for learning counting number, addition, subtraction etc. and leaves are used as vegetables while playing kitchen-set.
84	*Tectona grandis* L.f.	Verbenaceae	Teak	Sagun	Leaf	Used for colouration of palms by rubbing
85	*Tinospora cordifolia* (Willd.) Miers.	Menispermaceae	Indian tinospora	Gulbel	Stem	Used as telephone cord
86	*Trewia nudiflora* L.	Euphorbiaceae	False white teak	Pindar	Fruit	Used for making Wheels as play material

S.N	Name of the Plant	Family	Common Name	Hindi Name	Parts Used As Play Material	Mode of use
87	*Tridax procumbens* (L.)	Asteraceae	Coat buttons	Khal-muriya	Flower	Used to make tiara
88	*Triumfetta rhomboidea* Jacq.	Tiliaceae	Burr Bush	Chikti	Fruit	Used for making fun by throwing it to the hair of each other.
89	*Quisqualis indica* L.	Combretaceae	Rangoon creeper	Madhumalti	Flower	Used for making Ornaments (necklace, ring)
90	*Xanthium strumarium* L.	Asteraceae	Common Cocklebur	Chhota dhatura	Fruit	Used for throwing at each other and sticking it to their hairs and dresses